IDENTIFICATION OF PATHOGENIC FUNGI

IDENTIFICATION OF PATHOGENIC FUNGI

Colin K Campbell BSc MSc PhD
PHLS Mycology Reference Laboratory, Bristol Public Health Laboratory, UK

Elizabeth M Johnson BSc PhD
PHLS Mycology Reference Laboratory, Bristol Public Health Laboratory, UK

Christine M Philpot BSc PhD FRCPath
Cardiff Public Health Laboratory, UK

David W Warnock BSc PhD FRCPath
PHLS Mycology Reference Laboratory, Bristol Public Health Laboratory, UK

LONDON – PUBLIC HEALTH LABORATORY SERVICE

Identification of Pathogenic Fungi

© Public Health Laboratory Service 1996

All rights reserved. No part of this publication may be reproduced, stored in a retrieval system, or transmitted in any form or by any means, electronic, mechanical, photocopying, recording or otherwise, without the written prior permission of the publisher.

Published by the
Public Health Laboratory Service
61 Colindale Avenue
London NW9 5DF

ISBN 0 901144 39 8

Printed in England by Butler and Tanner, Frome, Somerset

CONTENTS

Preface   1

1 Introduction   2

2 Identification of moulds   10

3 Moulds with arthrospores   16
   Scytalidium dimidiatum   18
   Scytalidium hyalinum   20
   Coccidioides immitis   22
   Onychocola canadensis   24

4 Moulds with aleuriospores I.
   The dermatophytes   26
   Microsporum canis   32
   Microsporum equinum   34
   Microsporum gypseum   36
   Microsporum fulvum   37
   Epidermophyton floccosum   38
   Trichophyton terrestre   40
   Microsporum persicolor   42
   Trichophyton equinum   44
   Trichophyton erinacei   46
   Trichophyton mentagrophytes   48
   Trichophyton interdigitale   50
   Trichophyton rubrum   52
   Trichophyton rubrum (granular form)   53
   Trichophyton tonsurans   54
   Trichophyton soudanense   56
   Microsporum audouinii   58
   Trichophyton schoenleinii   60
   Trichophyton verrucosum   62
   Trichophyton violaceum   64
   Trichophyton concentricum   66
   Other *Microsporum* and
      Trichophyton species   68

5 Moulds with aleuriospores II.
   Others   72
   Geomyces pannorum   74
   Chrysosporium keratinophilum   76
   Myceliophthora thermophila   78
   Histoplasma capsulatum   80
   Blastomyces dermatitidis   82
   Paracoccidioides brasiliensis   84

6 Moulds with holoblastic conidia   86
   Aureobasidium pullulans   90
   Sporothrix schenckii   92
   Cladophialophora carrionii   94
   Cladophialophora bantiana   95
   Cladosporium sphaerospermum   96
   Cladosporium herbarum   97
   Cladosporium cladosporioides   97
   Fonsecaea pedrosoi   98
   Rhinocladiella atrovirens   100
   Ramichloridium mackenziei   102
   Ochroconis gallopava   104
   Alternaria alternata   106
   Ulocladium chartarum   108
   Curvularia lunata   110
   Bipolaris hawaiiensis   112
   Bipolaris australiensis   113
   Exserohilum rostratum   114
   Exserohilum longirostratum   115
   Exserohilum mcginnisii   115

7 Moulds with enteroblastic
   conidia adhering in chains   116
   Aspergillus flavus   120
   Aspergillus fumigatus   122
   Aspergillus glaucus   124
   Aspergillus nidulans   126
   Aspergillus versicolor   128
   Aspergillus ustus   130
   Aspergillus niger   132
   Aspergillus terreus   134
   Aspergillus candidus   136
   Penicillium marneffei   138
   Scopulariopsis brevicaulis   140
   Paecilomyces lilacinus   142
   Paecilomyces variotii   144

8	**Moulds with enteroblastic conidia adhering in wet masses**	**146**
	Cylindrocarpon lichenicola	150
	Fusarium dimerum	152
	Fusarium semitectum	154
	Fusarium moniliforme	156
	Fusarium oxysporum	158
	Fusarium solani	160
	Acremonium strictum	162
	Acremonium kiliense	164
	Lecythophora mutabilis	166
	Lecythophora hoffmannii	167
	Phialemonium spp.	167
	Scedosporium prolificans	168
	Scedosporium apiospermum	170
	Phialophora parasitica	172
	Phialophora richardsiae	174
	Phialophora verrucosa	176
	Phaeoannellomyces werneckii	178
	Exophiala spinifera	180
	Exophiala dermatitidis	182
	Exophiala jeanselmei	184
9	**Mucoraceous moulds and their relatives**	**186**
	Cunninghamella bertholletiae	190
	Absidia corymbifera	192
	Rhizomucor pusillus	194
	Mucor circinelloides	196
	Rhizopus microsporus	198
	Rhizopus arrhizus	200
	Rhizopus stolonifer	201
	Mucor hiemalis	202
	Mucor racemosus	203
	Basidiobolus ranarum	204
	Conidiobolus coronatus	206
	Pythium insidiosum	208
	Apophysomyces elegans	210
	Saksenaea vasiformis	212
	Mortierella wolfii	214
10	**Miscellaneous moulds**	**216**
	Aphanoascus fulvescens	218
	Monascus ruber	220
	Chaetomium globosum	222
	Phoma herbarum	224
	Myxotrichum deflexum	226
	Schizophyllum commune	228
	Leptosphaeria senegalensis	230
	Neotestudina rosatii	232
	Piedraia hortae	234
	Lasiodiplodia theobromae	236
	Pyrenochaeta romeroi	238
	Pyrenochaeta unguis-hominis	239
	Madurella mycetomatis	240
	Madurella grisea	241
11	**Identification of yeasts**	**242**
	Candida albicans	250
	Candida tropicalis	252
	Candida krusei	254
	Candida lipolytica	256
	Candida kefyr	258
	Candida lusitaniae	260
	Candida parapsilosis	262
	Candida pelliculosa	264
	Candida glabrata	266
	Candida guilliermondii	268
	Cryptococcus neoformans	270
	Rhodotorula glutinis	272
	Saccharomyces cerevisiae	274
	Geotrichum candidum	276
	Blastoschizomyces capitatus	278
	Trichosporon beigelii	280
	Malassezia furfur	282
	Malassezia pachydermatis	284
Appendices		
1	Common mycological terms	286
2	Further reading	290
Index		**292**

PREFACE

There is widespread agreement that opportunistic fungal infections, such as aspergillosis and candidosis, are no longer uncommon disorders. These infections are occurring in ever-increasing numbers among neutropenic cancer patients, transplant recipients and patients with AIDS. In addition to the rise in the number of infections due to such well-recognised pathogens as *Aspergillus fumigatus*, an increasing number of ubiquitous environmental moulds are being implicated as the cause of significant human infection. These moulds are organisms whose natural habitat is in the soil or on plants, compost heaps or rotting food. Many are familiar to mycologists, plant pathologists and food microbiologists, but they present problems for the clinical microbiologist, who often has had no formal training in the identification of fungi. It is becoming more important for the clinical laboratory to be able to recognise these organisms because many do not respond to standard treatment.

This manual has been designed for use by medical, scientific and technical staff in hospital laboratories in the UK and abroad, but we hope it will also be of interest to scientists in other institutions. The organisms described have been grouped in chapters according to spore-bearing structures produced in culture, rather than ordered on an alphabetical basis. Each chapter has been arranged so that the descriptions for similar organisms may be found on adjacent pages. In addition, we have attempted to provide differential diagnoses on the basis of both colonial appearance and microscopic characteristics for the organisms described. Lack of space has precluded the inclusion of every rare organism that might be isolated from a clinical specimen. In some cases a single representative member of a genus is described, and isolates that appear similar to the description provided may need to be referred to a specialist for confirmation of the identification.

Two appendices have been included in this manual. The first gives definitions for some mycological terms in common use; the second lists some useful monographs and more comprehensive texts that the reader may wish to consult. A detailed index will be found at the end of this monograph.

Before she died in October 1995, Christine Philpot had made a significant contribution to our manuscript. We hope the finished monograph will be a fitting memorial to her.

We thank Nina Hukkanen and Rodney Wilson of PHLS Publications, Headquarters, Public Health Laboratory Service, for their invaluable help in the design and production of this monograph.

Colin K Campbell
Elizabeth M Johnson
David W Warnock

1 INTRODUCTION

The Kingdom Fungi consists of a distinct group of eukaryotic organisms that absorb their nourishment from living or dead organisms or organic matter. Fungi are found throughout nature, performing an essential service in returning to the soil nutrients removed by plants. There is, however, a large group of species that are parasitic on plants and a smaller group that are parasitic on animals, as well as on man.

In most fungi, the vegetative stage consists of a system of branching filaments, or *mycelium*. Each individual filament, or *hypha*, has a rigid cell wall and increases in length as a result of apical growth. In the more primitive fungi, the mycelium remains *aseptate* (without cross-walls). In the more advanced groups, however, the mycelium is *septate* with more or less frequent cross-walls. The individual reproductive bodies of fungi, or *spores*, consist of a single cell or several cells contained within a rigid wall. During their evolution most fungi have relied upon a combination of sexual and asexual reproductive mechanisms to assist their survival. Their sexual spores and the protective structures that develop around them form the main basis for fungal classification.

Yeasts are unicellular fungi consisting of separate, round, oval or elongated cells, or *blastospores*, that propagate by budding out similar cells from their surface. The bud may become detached from the parent cell, or it may remain attached and itself produce another bud. In this way a chain of cells may be produced. Under certain conditions, continued elongation of the parent cell before it buds results in a chain of elongated cells, or *pseudohypha*. Some yeasts can also produce true hyphae, with cross-walls.

Some species of fungi are homothallic and able to form sexual structures within individual colonies. Most, however, are heterothallic and do not form their sexual structures unless two different mating strains come into contact. Thus, sexual reproduction is often difficult to obtain in culture, and this helps to explain the large number of fungi that bear two names, one designating their sexual stage (or *teleomorph*) and the other their asexual stage (or *anamorph*). Both these names are valid under the International Code of Botanical Nomenclature, but that of the teleomorph should take precedence over that of the anamorph. In practice, however, it is more common to refer to moulds by their asexual designation because this is the stage which is usually obtained in culture.

SEXUAL REPRODUCTION

The Fungi are classified into three main divisions – the *Zygomycotina*, the *Ascomycotina* and the *Basidiomycotina* – according to their method of sexual reproduction. In the division Zygomycotina, fusion of the tips of two hyphae leads to the formation of a single, large *zygospore* between them. This is a multinucleate thick-walled structure that has evolved to endure adverse environmental conditions. Meiosis occurs on germination and the vegetative haploid mycelium develops. In contrast, in the divisions Ascomycotina and Basidiomycotina, sexual reproduction has evolved into a means of rapid dispersal to new habitats, unlike the resting nature of the zygospore. In both these groups the diploid stage is transient, with meiosis resulting in the production of enormous numbers of short-lived haploid spores.

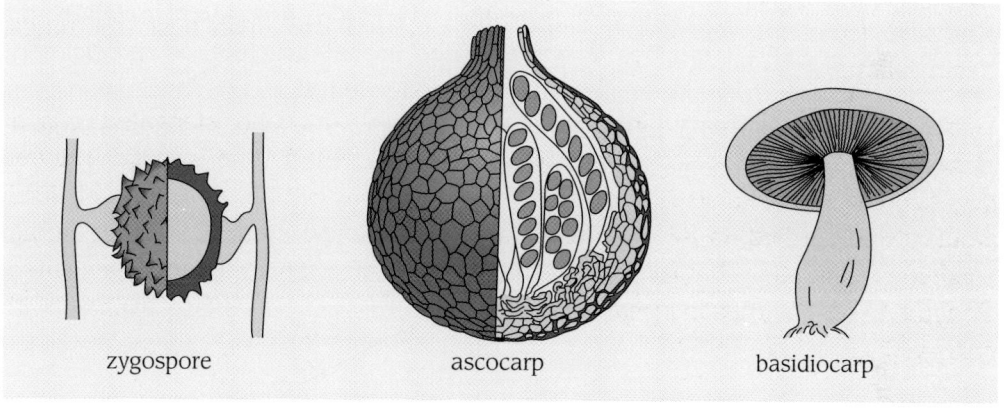

zygospore   ascocarp   basidiocarp

In the Ascomycotina, the sexual spores, or *ascospores*, are produced in sacs, or *asci*. Each ascus usually contains eight ascospores. The group shows a gradual transition from primitive forms that produce single asci to species that produce large structures, or *ascocarps*, containing large numbers of asci. Three main forms of ascocarp are common: the *perithecium*, which releases its spores through an apical opening; the *cleistothecium*, which splits open to liberate its contents; and the *gymnothecium*, which is an open loose network of protective hyphae. In most of the Basidiomycotina, the sexual spores, or *basidiospores*, are borne on projections at the tip of club-shaped *basidia*. These are produced in macroscopic structures, or *basidiocarps*. However, most of the organisms described in this manual are identified on the basis of their asexual reproductive structures and spores.

ASEXUAL REPRODUCTION

Fungi belonging to all three divisions also produce asexual spores by simple haploid nuclear division. Again, short-lived propagules are produced in enormous numbers to ensure spread to new habitats. In many fungi this asexual (or imperfect) stage has proved so successful that the sexual (or perfect) stage has diminished or even disappeared. These are the so-called *Fungi Imperfecti* which are grouped into an artificial division, the *Deuteromycotina*. This division by convention contains all the asexual relatives of the Ascomycotina and Basidiomycotina, but not those of the Zygomycotina. Yeasts are found in three of the four divisions of the Fungi: the Ascomycotina, the Basidiomycotina and the Deuteromycotina.

In the Ascomycotina, Basidiomycotina and Deuteromycotina the asexual spores are termed *conidia*, and are produced from a *conidiogenous cell*. In some species the conidiogenous cell is not different from the rest of the mycelium. In others the conidiogenous cell is contained in a specialised hyphal structure, or *conidiophore*. There are two basic methods of asexual spore production: *thallic*, in which an existing hyphal cell is converted into a conidium; and *blastic*, in which the conidium is produced as a result of some form of budding process.

THALLIC CONIDIOGENESIS

In thallic conidiogenesis the conidium is produced from an existing hyphal cell. This occurs when a hypha breaks up into sections to form individual cells, or *arthrospores*, or when one cell develops a thick wall to form a resting spore, or *chlamydospore*.

Arthrospores are derived from the fragmentation of an existing hypha and represent the simplest form of asexual sporulation. In most species the septum separating two cells splits down the middle, leaving a trace of the resulting torn wall on the end of the spore. In a few instances the arthrospores are intercalated with separating cells and are liberated after these cells have dissolved. This leaves a marked annular frill at the ends of the detached arthrospores. Moulds which produce arthrospores as their principal reproductive spores are described in detail in Chapter 3.

Aleuriospores represent an intermediate state between thallic and blastic conidiogenesis. These spores are formed from the side or tip of a hypha and, during the initial stage before a septum is laid down, can resemble short, hyphal branches. As in all genuine cases of thallic conidiogenesis, it is not possible for a second spore to be formed at the same point. This form of conidium is characteristic of the dermatophytes (described in Chapter 4), but is also found in a number of other fungi of medical importance (described in Chapter 5).

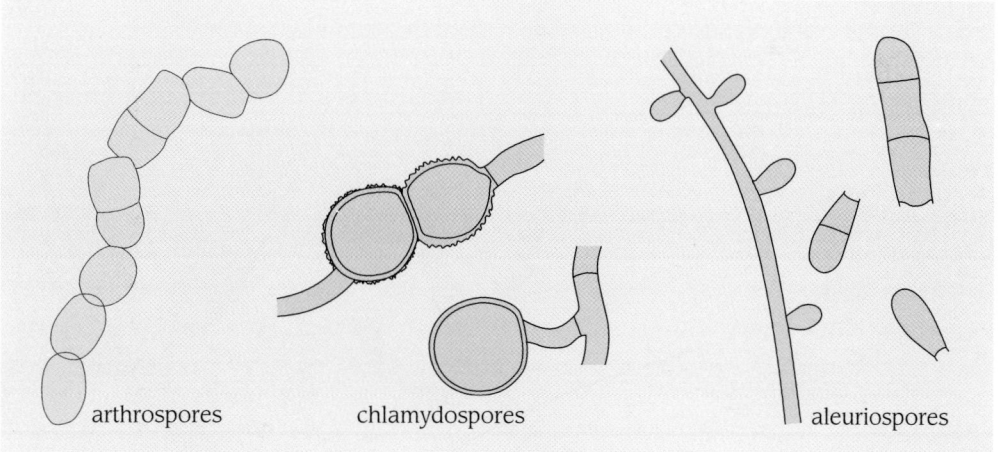

arthrospores   chlamydospores   aleuriospores

BLASTIC CONIDIOGENESIS

Many fungi have evolved some form of repeated budding that permits them to produce large numbers of asexual spores from a single conidiogenous cell. Two forms of blastic conidiogenesis are now recognised: *holoblastic development*, in which both the inner and the outer wall of the conidiogenous cell swell out to form the conidium, and *enteroblastic development*, in which the conidium is produced from within the conidiogenous cell, the outer layer of the hyphal wall being ruptured and an inner layer extending through to become the new spore wall. These two forms of blastic conidiogenesis can be further subdivided according to the details of spore development.

HOLOBLASTIC CONIDIOGENESIS

In some fungi, the conidiogenous cells each produce a single holoblastic conidium. In others, however, the first-formed conidium produces a second conidium and the second produces a third, and so on, until a chain of spores is produced with the youngest at its tip. As each conidium can produce more than one bud, a branching chain becomes possible. Examples of moulds that produce holoblastic branching chains of spores include species of *Cladosporium*. In other species, the conidiogenous cell that produced the first-formed spore then grows past it to produce a second. If this process is repeated, it will result in a elongated conidiogenous cell with numerous lateral single spores along its sides. This happens in species of *Ulocladium*. Moulds which produce holoblastic conidia are described in detail in Chapter 6.

Holoblastic conidia

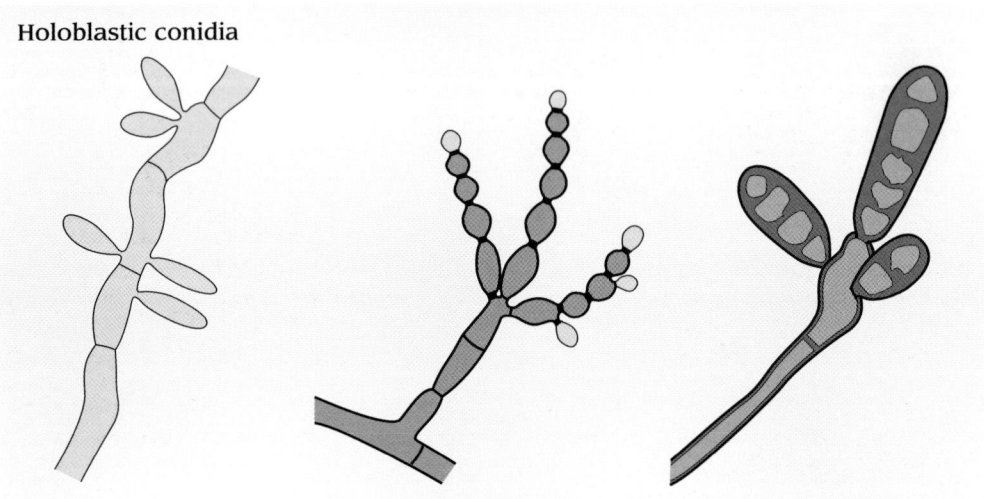

ENTEROBLASTIC CONIDIOGENESIS

In fungi that produce enteroblastic spores, the wall of the conidia is derived from the inner layer of the wall of the conidiogenous cell and the conidia are produced from an opening in the outer wall of the conidiogenous cell. This permits a succession of spores to be produced at the same point. The specialised conidiogenous cell from which the conidia are produced is termed a *phialide*. In some fungi, such as species of *Aspergillus* and *Penicillium*, continuous replenishment of the inner wall of the tip of the phialide results in the formation of an unbranched chain of connected spores, with the youngest at the base. Moulds which produce enteroblastic conidia in chains are described in detail in Chapter 7.

In other fungi, such as species of *Fusarium* and *Acremonium*, a new inner layer of wall material is produced for each successive spore. Repeated conidiogenesis results in an accumulation of the unused remains of these layers within the tip of the phialide. The spores are not firmly attached to each other and often move aside to accumulate in a wet mass around the phialide. Unlike the spores of species of *Aspergillus* and *Penicillium*, these spores do not spread on air currents, but are coated with a wettable slime which appears to be an adaptation to water dispersal. Moulds which produce enteroblastic conidia in wet masses are described in detail in Chapter 8.

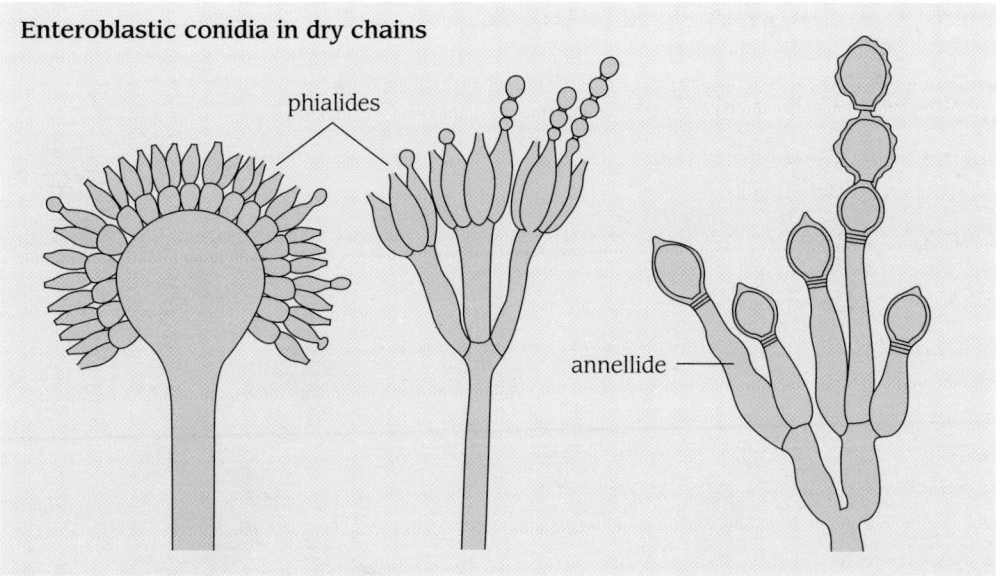

Enteroblastic conidia in dry chains
phialides
annellide

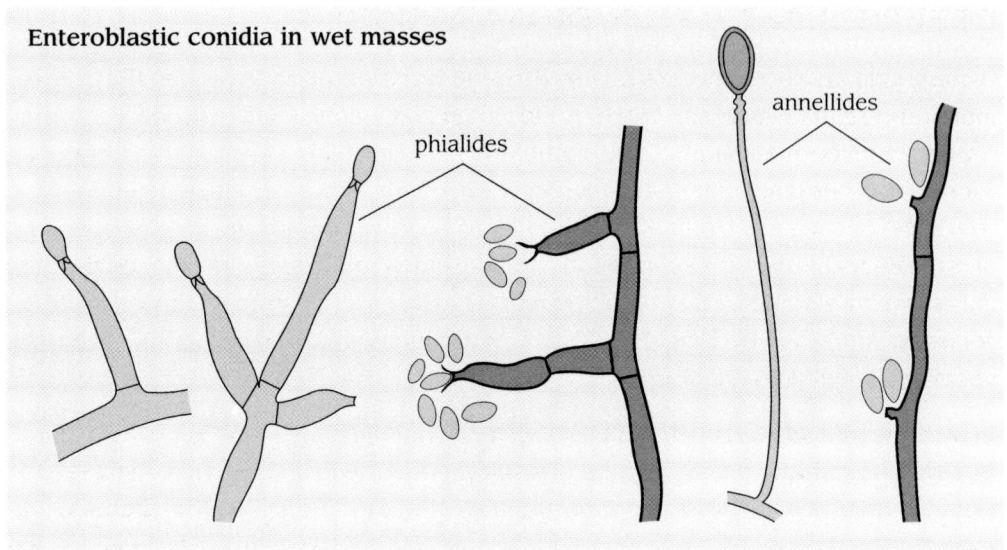

Enteroblastic conidia in wet masses

Annellides, like phialides, are cells which produce conidia at their tips in unbranched chains (as in the genus *Scopulariopsis*) or in wet masses (as in the genus *Scedosporium*). Unlike phialides, annellides increase in length each time a new spore is produced. An old annellide that has produced many spores will have a number of apical scars or annellations at its tip. These scars, which are left as successive spores break off, are often difficult to see under the optical microscope.

OTHER FORMS OF SPORE PRODUCTION

In the Zygomycotina, one major group the Order Mucorales produces the asexual spores, or *sporangiospores*, inside a closed sac, or *sporangium*, the wall of which ruptures to liberate them. The sporangium is held above the substratum on an unbranched or branched *sporangiophore*. The different species of fungi in this group are distinguished from one another by the sporangiophores, sporangia and sporangiospores, as well as by the presence or absence of *rhizoids* that anchor the sporangiophores to the substratum. In addition these organisms have large, aseptate or almost aseptate hyphae. These and other members of the Zygomycotina are described in detail in Chapter 9.

Most of the moulds described in this manual are identified on the basis of their asexual reproductive structures and spores. However, there are a number of pathogenic moulds that produce sexual spores in ascocarps or basidiocarps, rather than asexual spores, in culture and these are described in detail in Chapter 10. This chapter also includes descriptions of several moulds that produce macroscopic fruiting bodies (*pycnidia*) containing asexual spores. In addition, several non-sporing pathogenic moulds have been included.

The identification of yeasts, unlike that of moulds, relies on a combination of morphological and biochemical characteristics. Chapter 11 deals with the organisms that are most frequently encountered in clinical laboratories and describes the tests that are most commonly employed for their identification.

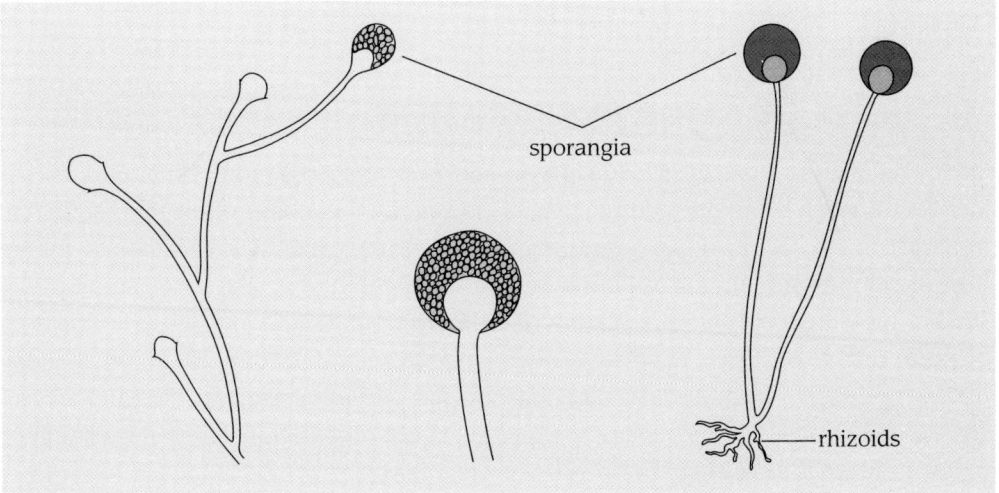

2 IDENTIFICATION OF MOULDS

The identification of filamentous fungi is based on the examination of their macroscopic (colonial) and microscopic characteristics. Macroscopic features such as colonial form, surface colour and production of pigments are often helpful in identification. The growth rate of mould colonies depends on the culture medium and temperature of incubation but, provided conditions are standardised, these characteristics can be taken into consideration in the process of identification. Morphological examination of microscopic structures such as spores and spore-bearing cells is an essential part of mould identification. Moulds that fail to sporulate are often impossible to speciate and it is therefore important to select culture conditions which favour sporulation.

MEDIA

The texture and colour of mould colonies often depend on the age of the culture and the agar medium on which the organism is grown. Nevertheless, these characteristics are useful in identification. Owing to the almost universal use of Sabouraud's glucose peptone agar, the descriptions in this manual are based on cultures prepared on it. However, there are numerous formulations of that medium, both with and without antibiotics, and it is advisable to confine supplies to one manufacturer as the morphological appearance of moulds, and pigmentation in particular, can differ from one formulation to another. Moulds often grow best on rich media, such as glucose peptone agar, but over-production of mycelium often results in loss of sporulation. If a mould isolate fails to produce spores or other recognisable structures after two weeks, it should be subcultured to a less-rich medium to encourage sporulation and permit identification. The composition of a number of useful media is given at the end of this chapter.

METHODS OF SLIDE PREPARATION

Microscopic examination of slide preparations is the most important part of the identification of a mould culture. If well prepared, these will often give sufficient information on the form and arrangement of spores and other structures for an identification of the fungus to be made. The usual method is to remove some of the surface growth from a culture plate with a sharp rigid needle and place it in a drop of mounting fluid (such as lactofuchsin or lactophenol cotton blue) on a clean microscope slide. The material is then teased apart with two sharp needles and a cover slip applied. Gentle pressure is used to spread out the preparation before it is examined under a microscope using ×10 and ×40 objective lenses.

There are several other methods for the preparation of slides for microscopic examination of fungi. One of the most helpful is to use clear adhesive tape. A small 'flag' of tape (about 20 mm long) is cut with scissors and placed on the end of a rigid needle. The tape is pressed, adhesive side downwards, on to the surface of the culture using a second needle applied to the back of the tape. The coated tape is then placed, adhesive side upwards, in a small drop of mounting fluid on a microscope slide. A second small drop of mounting fluid is placed on the preparation and a cover slip applied.

If a slide preparation shows no spores, it is often helpful to try nearer the centre of the colonies, where the mould is older and has had more time to sporulate. If there are too many spores and the sporing structures cannot be discerned, it is useful to try nearer the edge of the colonies. If no spores are found in slide preparations, it is sometimes worthwhile to remove the lid from the culture plate and examine the colonies for evidence of sporulation under the low-power objective of a microscope.

Both the 'needle' and the 'tape' methods give suitable preparations for microscopic examination, but each has its drawbacks with certain forms of fungal growth. Needle preparations dislodge chains and wet masses of spores, and these features are best seen with tape preparations. On the other hand, structures such as pycnidia and spores hidden deep in the mycelium are not picked up on tape and require dissection with a needle. The needle also allows sub-agar growth to be studied. Mounting fluids such as lactophenol, but not lactofuchsin, attack adhesive tape and render it unsuitable for preparation of permanent mounts. Needle preparations can be sealed around the edge with DPX for long-term storage.

SLIDE CULTURE

The slide culture technique is useful for observing the intact arrangement of spores or spore-bearing structures. A thin, square block of a suitable nutrient agar (smaller than a cover slip) is placed on a sterile microscope slide supported on a bent glass rod in a petri dish. The four sides of the agar block are then inoculated with portions of mycelium of the fungus to be identified. The block is then covered with a sterile cover slip, sterile distilled water added to the base of the petri dish, the lid replaced and the plate incubated at 30°C. Once adequate sporulation has occurred, the cover slip is removed from the agar and placed on a drop of mounting fluid on a clean glass slide with the adherent mycelium downwards. The agar block is then removed and discarded, leaving adherent mycelium on the slide. Mounting fluid is added and a clean cover slip applied. The preparations can be sealed for long-term preservation.

MEDIA FOR MOULD IDENTIFICATION

Cornmeal agar

This medium is useful for stimulating ascocarp and pycnidium production in some moulds.

cornmeal extract	2 g
agar	15 g
distilled water	1 L

Heat to dissolve. Autoclave at 121°C for 15 min.

Czapek-Dox agar

This defined medium is recommended for the identification of *Aspergillus* and *Penicillium* spp. It is also useful for stimulating sporangium production in mucoraceous moulds.

sucrose	30 g
sodium nitrate	2 g
potassium chloride	0.5 g
magnesium glycerophosphate	0.5 g
potassium sulphate	0.35 g
ferrous sulphate	0.01 g
agar	12 g
distilled water	1 L

Heat to dissolve. Autoclave at 121°C for 15 min.

Dermatophyte test agar

This medium changes turns red in colour with dermatophytes and is useful for distinguishing those species from other moulds. It is important to remember that some non-dermatophyte moulds can also produce that colour change.

glucose	40 g
mycological peptone	10 g
phenol red	0.2 g
agar	12 g
distilled water	1 L

Heat to dissolve. Autoclave at 121°C for 15 min.

Malt extract agar

This rich medium is recommended as an alternative to Sabouraud's glucose peptone agar for stimulating sporulation in a wide range of moulds, including the dermatophytes.

malt extract	30 g
mycological peptone	5 g
agar	15 g
distilled water	1 L

Heat to dissolve. Autoclave at 115°C for 10 min.

Philpot's urea agar

This medium is used to distinguish *Trichophyton rubrum* (urease negative) from *T. interdigitale* (urease positive). It is important to remember that the granular form of *T. rubrum* gives a positive result, as will most dermatophytes.

glucose	5 g
mycological peptone	1 g
sodium chloride	5 g
potassium dihydrogen orthophosphate	2 g
phenol red	0.012 g
agar	15 g
distilled water	1 L

Heat to dissolve. Autoclave at 115°C for 20 min. Cool to 50°C and add 50 mL of sterile 40% urea solution.

Potato dextrose agar

This is a good general-purpose medium which stimulates sporulation in many moulds. It stimulates pigment production in some dermatophytes.

glucose	20 g
potato extract	4 g
agar	15 g
distilled water	1 L

Heat to dissolve. Autoclave at 121°C for 15 min.

Glucose peptone agar

This medium is recommended for the isolation and cultivation of dermatophytes and other moulds requiring a rich substrate with a high content of organic nitrogen. Antibacterial antibiotics (in particular chloramphenicol) can be added to control bacterial contamination.

glucose	40 g
mycological peptone	10 g
agar	15 g
distilled water	1 L

Heat to dissolve. Autoclave at 121°C for 15 min.

MOUNTING FLUIDS

Lactophenol

phenol crystals	20 g
lactic acid	20 mL
glycerol	40 mL
distilled water	20 mL

Heat gently to dissolve. Store away from direct sunlight.

Lactophenol cotton blue

cotton blue	0.075 g
lactophenol	100 mL

Store away from direct sunlight.

Lactofuchsin

acid fuchsin	0.1 g
lactic acid	100 mL

Store away from direct sunlight.

3 MOULDS WITH ARTHROSPORES

INTRODUCTION

Arthrospores, which are derived from the fragmentation of existing hyphae, represent the simplest form of sporulation and have evolved in several fungal groups. The clinical laboratory will encounter many arthrosporic moulds, many only as contaminants. In this chapter only those moulds of clinical significance in which arthrospores are the primary distinctive structures are described. Moulds that produce arthrospores secondarily to another spore type, e.g. the dermatophytes, are treated in later chapters, as are those fungi with a predominantly yeast-like colonial form, e.g. *Geotrichum candidum*. A third group, excluded here, are pure white, fast-growing, floccose moulds many of which develop as contaminants from airborne basidiomycete spores.

Arthrospores are generally absent from colonies pigmented brown, green, red or purple. Most arthrosporic fungi lack pigmentation and give white or cream colonies. The notable exception to this is *Scytalidium dimidiatum*, with brown or black colonies. The first step in the examination of arthrosporic moulds should be to ascertain whether another spore-form is present. If so, identification should be based upon that form. If not, the method of separation of the arthrospores from the parent hyphae should be sought. In most types of fungi the septum separating two cells splits down the middle, leaving a trace of the resulting torn wall on the end of the cylindrical spore. In a few instances, notably the Hazard Group 3 organism *Coccidioides immitis*, arthrospores are released by the dissolution of a specialised empty cell at either side. This leaves a marked frill at the ends representing the remnants of the adjacent fractured cells.

Key to arthrosporic moulds

1a	Colony dark brown or black	*Scytalidium dimidiatum*
1b	Colony white or cream	2
2a	Arthrospores regularly alternating with empty cells	3
2b	Arthrospores not separated by empty cells	4
3a	Arthrospores mostly wider than 2 µm (see also some dermatophytes in Chapter 4)	*Coccidioides immitis*
3b	Arthrospores mostly 1-1.5 µm	*Malbranchea* spp.
4a	Colony with little aerial mycelium (see Chapter 11)	*Geotrichum* spp. or *Trichosporon* spp.
4b	Colony floccose	5
5a	Growth fast (diameter > 20 mm in 1 week)	6
5b	Growth slow	*Onychocola canadensis*
6a	Arthrospores abundant, 4-8 µm wide, becoming rounded, often with a central septum	*Scytalidium hyalinum*
6b	Arthrospores few, 2-5 µm wide, flat ended	basidiomycete moulds

SCYTALIDIUM DIMIDIATUM

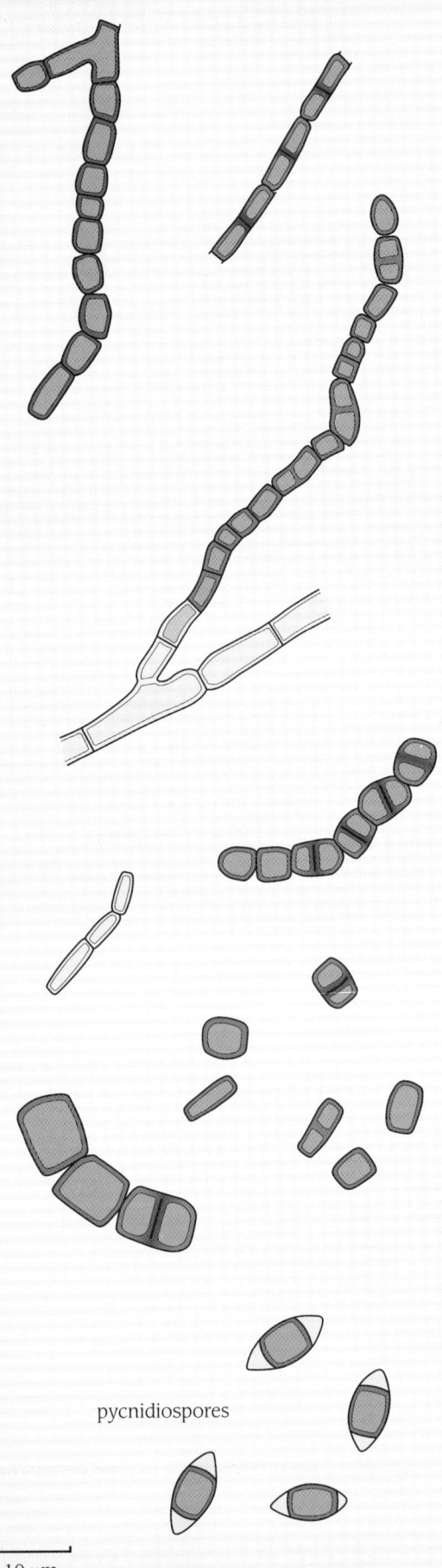

pycnidiospores

10 μm

COLONIAL APPEARANCE
at 30°C on glucose peptone agar

diameter	90 mm in three days in most strains (but see variant forms)
topography	abundant aerial growth to lid of petri dish
texture	floccose
colour	white at first, soon becoming black or dark brown
reverse	black or dark brown

MICROSCOPIC APPEARANCE
at 30°C

predominant features	brown arthrospores
arthrospores	narrow, colourless arthrospores and wider, brown-walled arthrospores are produced in abundance on the aerial mycelium; many have two cells separated by a thick septum
pycnidia	hard, black stromata about 1-2 mm across are formed on the surface of old cultures in some strains. When dissected these show multiple pycnidial cavities filled with pycnidiospores. The latter are unicellular and colourless when immature, but become three-celled with the central cell darker than the end cells. The pycnidial state is referred to *Nattrassia mangiferae*

VARIANT FORMS

Very slow growing forms (diameter 10 mm in one week) with compact, velvety olive-brown colonies with black, submerged edge and few arthrospores have been described, mainly from infections originating in the Indian subcontinent.

DIFFERENTIAL DIAGNOSIS

colonial appearance *Lasiodiplodia theobromae* and some other moulds, which lack the characteristic arthrospores

microscopic appearance *Scytalidium hyalinum* has similar arthrospores, but lacks the brown pigment

SEXUAL STATE

None known.

CLINICAL IMPORTANCE

It is a well-recognised cause of skin and nail infection of the hands and feet in tropical countries.

SCYTALIDIUM HYALINUM

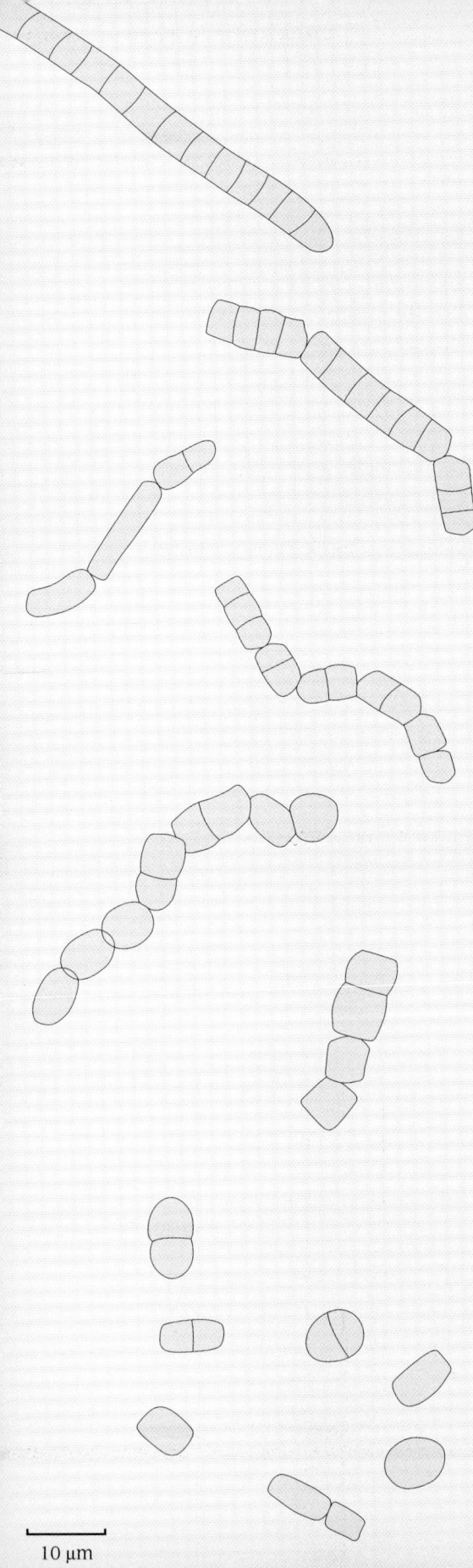

10 µm

COLONIAL APPEARANCE
at 30°C on glucose peptone agar

diameter	90 mm in three days
topography	abundant aerial growth to lid of petri dish
texture	floccose
colour	white to pale cream
reverse	colourless to pale brown

MICROSCOPIC APPEARANCE
at 30°C

predominant features	colourless arthrospores
arthrospores	vary in shape from elongate narrow forms to short, wide cylinders which tend to round off in old cultures to become almost spherical; as with *S. dimidiatum*, two-celled forms are frequent

VARIANT FORMS

Occasional strains with some degree of brown pigmentation suggest close affinities with *S. dimidiatum*.

DIFFERENTIAL DIAGNOSIS

colonial appearance	*Arthrinium* spp. and many basidiomycetous moulds; *Chrysosporium* spp. and some dermatophytes may resemble old colonies
microscopic appearance	*Geotrichum candidum* has large colourless arthrospores, but lacks the two-celled forms

SEXUAL STATE

None known.

CLINICAL IMPORTANCE

It is a cause of skin and nail infection of the hands and feet in tropical countries.

COCCIDIOIDES IMMITIS

(Hazard Group 3 pathogen)

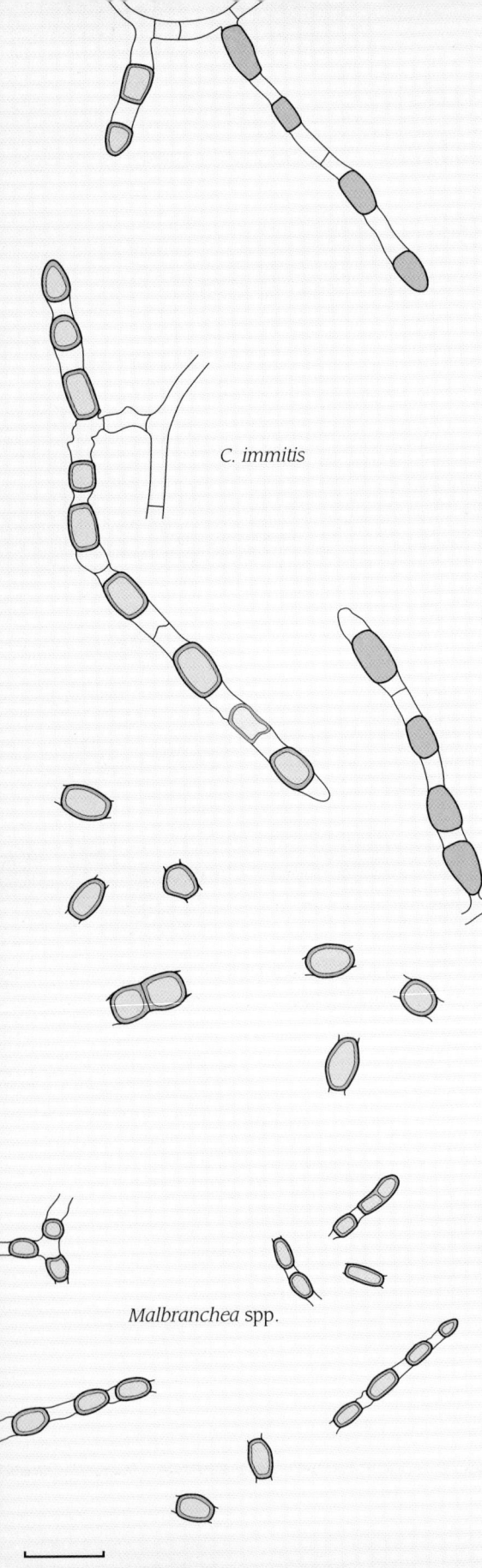

C. immitis

Malbranchea spp.

10 μm

COLONIAL APPEARANCE
at 30°C on glucose peptone agar

diameter	10-20 mm in one week
topography	flat, with entire or irregular margin
texture	glabrous at first, but soon becoming floccose
colour	pale to medium grey
reverse	colourless to brown

MICROSCOPIC APPEARANCE
at 30°C

predominant features	many small, barrel-shaped arthrospores with prominent end scars
arthrospores	thick-walled, barrel-shaped, mostly 2.5-4.5 μm × 3-8 μm; borne on terminal hyphae branching at right angles and alternating with thin-walled cells that are devoid of contents and that rupture to release the adjacent spores

DIFFERENTIAL DIAGNOSIS

colonial appearance many dermatophytes, *Chrysosporium* spp., *Blastomyces dermatitidis*, *Paracoccidioides brasiliensis* (like *C. immitis* all grow on cycloheximide-containing media)

microscopic appearance *Malbranchea* spp. are almost identical, but their arthrospores are often narrower and mostly cylindrical, rather than barrel-shaped

some species of dermatophytes and *Chrysosporium* produce less regular series of alternating arthrospores

Note: full identification of *C. immitis* requires demonstration of the appropriate exoantigen or conversion to the spherule form at 40°C on special media.

SEXUAL STATE

None known.

CLINICAL IMPORTANCE

It is the cause of coccidioidomycosis in man and mammals. Inhalation can bring about a transient pulmonary infection in normal individuals, which can proceed to cause progressive infection of the lungs or more generalised infection. This is lethal if left untreated. Most cases occur in the south-western USA and parts of Central and South America. This organism poses a serious threat to laboratory workers handling live cultures.

ONYCHOCOLA CANADENSIS

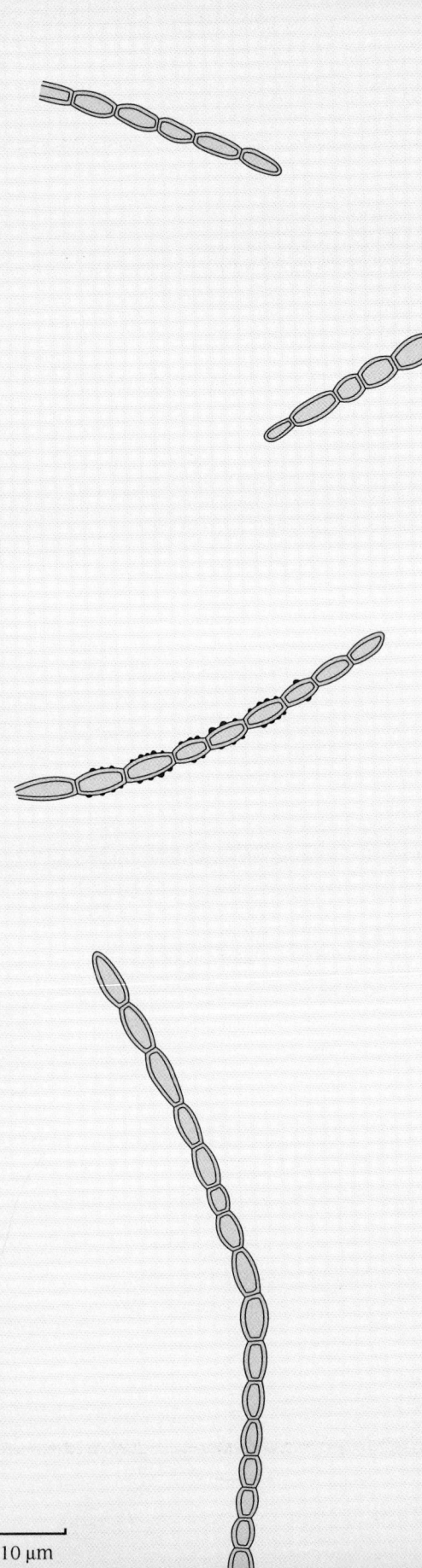

COLONIAL APPEARANCE
at 30°C on glucose peptone agar

diameter	5 mm in one week
topography	domed
texture	densely floccose to velvety
colour	white to cream or pale grey
reverse	pale brown to grey

MICROSCOPIC APPEARANCE
at 30°C

predominant features	long chains of small, oval arthrospores
arthrospores	slow to develop, especially on glucose peptone agar; borne on relatively undifferentiated hyphae; sometimes single and released by fracture of adjacent empty cells, or in long chains and then released (if at all) by splitting of the septa; individual spores ellipsoid to irregular in shape, smooth to finely rough, 4-8 µm × 2-5 µm, occasionally larger with two cells

DIFFERENTIAL DIAGNOSIS

colonial appearance *Chrysosporium* spp., some dermatophytes

microscopic appearance *Malbranchea* spp., *Coccidioides immitis*, many basidiomycetous moulds

SEXUAL STATE

Arachnomyces nodososetosus

CLINICAL IMPORTANCE

It is a rare cause of nail infection.

4 MOULDS WITH ALEURIOSPORES I

INTRODUCTION

The dermatophytes are specialised parasites of keratin and belong to the genera *Epidermophyton*, *Microsporum* and *Trichophyton*. Together with those fungi described in Chapter 5, they produce conidia known as aleuriospores in their saprophytic state (culture). Some species are found in soil (geophilic), others have one or more animal hosts (zoophilic) and others are confined to man (anthropophilic). Almost all may be causes of human infection, often in characteristic sites. Thus clinical details such as the site of infection and history of contact with animals or soil are of great help in laboratory identification.

The dermatophytes vary greatly in their rate of growth, colony topography, texture and colour and it is useful to examine these features before the microscopic appearance. In general, colonies range in colour from white, to cream or pink, to shades of red-brown or violet. Moulds that produce green, dark olive brown or black colonies are not dermatophytes.

Colonial characteristics are influenced by the agar on which the mould is grown. Owing to the almost universal use of Sabouraud's glucose peptone agar, the descriptions given in this chapter are based on that medium. Sporulation of dermatophytes is not always as good on glucose peptone agar as on malt agar and subculture may be necessary to establish an unidentifiable isolate as a dermatophyte by production of aleuriospores. It should be noted that the colonial characters and in some cases the microscopic features on these media can differ from those given in this chapter for glucose peptone agar.

To identify a dermatophyte, some of the surface growth should be removed from a glucose peptone agar plate and examined under a microscope to establish whether: (1) macroconidia are predominant; (2) microconidia are predominant; or (3) no spores are present. The keys on pages 29-31 permit the differentiation of the common dermatophytes to species level on the basis of a combination of microscopic and colonial characteristics.

Descriptions of a number of uncommon dermatophytes, most of which are soil saprophytes, are to be found at the end of this chapter. These species are not included in the keys.

It is important to remember that a number of related aleuriosporic fungi, in particular *Chrysosporium* species, can be mistaken for dermatophytes (see Chapter 5).

THE DERMATOPHYTES

NATURAL HABITATS AND SITES OF HUMAN INFECTION

The dermatophytes are described as geophilic, zoophilic or anthropophilic depending upon whether their normal habitat is the soil, an animal or humans (see Table 4.1). These ecological differences have important epidemiological implications in relation to the acquisition of human infection, the site(s) of infection, and the spread of infection between individuals.

Table 4.1 Habitat of dermatophytes and sites of human infection

Species		Habitat	Usual sites of infection
	Epidermophyton floccosum	man	groin, feet
***Microsporum* spp.**	*M. audouinii*	man	scalp
	M. canis	cat, dog	scalp, face, trunk, limbs
	M. equinum	horse	scalp, face, trunk, limbs
	M. fulvum	soil	face, trunk, limbs
	M. gypseum	soil	face, trunk, limbs
	M. persicolor	rodents	face, trunk, limbs
***Trichophyton* spp.**	*T. concentricum*	man	face, trunk, limbs
	T. erinacei	hedgehog	face, trunk, limbs
	T. equinum	horse	face, trunk, limbs
	T. interdigitale	man	feet, nails
	T. mentagrophytes	rodents	scalp, face, trunk, limbs
	T. rubrum	man	feet, groins, nails, trunk
	T. schoenleinii	man	scalp
	T. soudanense	man	scalp, face, trunk, limbs
	T. tonsurans	man	scalp, face, trunk, limbs
	T. verrucosum	cattle	scalp, face, trunk, limbs
	T. violaceum	man	scalp, face, trunk, limbs

HAIR INVASION BY DERMATOPHYTES

All dermatophytes have the same appearance in infected skin and nails, namely hyphae that are septate, regular in width, branched, and often fragmented into chains of arthrospores. In scalp and facial hair, the hyphae break up into arthrospores, the size and disposition of which gives some indication of the species of dermatophyte involved. (See Table 4.2) Arthrospores of *Microsporum canis* and *M. audouinii* are 2-3 μm in diameter and are formed in irregular clusters on the outside of hair shafts. This form of hair invasion is termed 'ectothrix'. Arthrospores of *Trichophyton* spp. and *M. gypseum* are larger and may be formed on the surface or inside the hair shafts ('ecto-endothrix'). Infection in which arthrospores are confined within the hair is termed 'endothrix'. In the condition known as favus, caused by *T. schoenleinii*, a distinctive form of endothrix hair invasion develops. Hyphae are found within the hairs which contain air spaces, but arthrospore formation does not occur. *Epidermophyton floccosum*, *Microsporum persicolor* and *Trichophyton concentricum* do not invade hair.

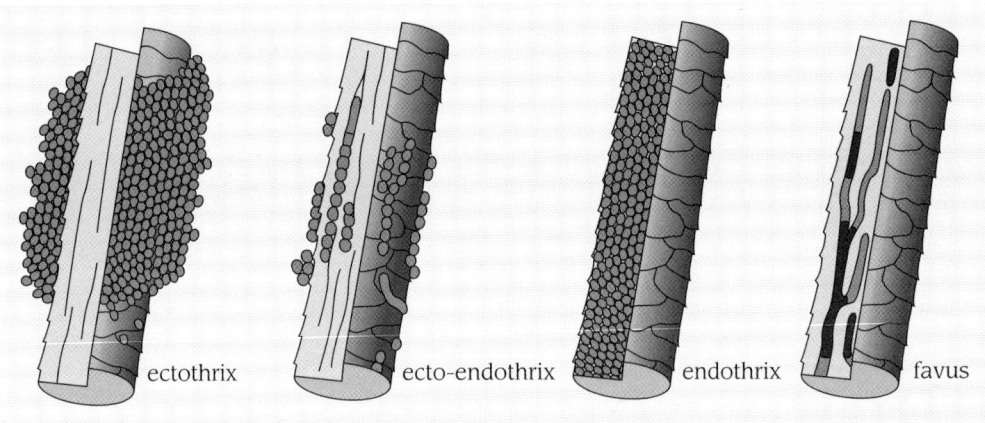

ectothrix   ecto-endothrix   endothrix   favus

Table 4.2 Characteristics of some common dermatophytes invading hair

Species	Arthrospore size (μm)		Arrangement
Microsporum audouinii	2-5	(small)	ectothrix
Microsporum canis	2-5	(small)	ectothrix
Trichophyton mentagrophytes	3-5	(small)	ectothrix
Trichophyton tonsurans	4-8	(large)	endothrix
Trichophyton verrucosum	5-10	(large)	ectothrix
Trichophyton violaceum	4-8	(large)	endothrix

Key to dermatophytes with macroconidia predominant

"beak" — rough walls

2a 2b 4a 4b 4c

1a	Macroconidia with some roughness on the outer surface		2
1b	Macroconidia smooth		4
2a	Macroconidia with a terminal 'beak'		3
2b	Macroconidia without a terminal 'beak'		*Microsporum gypseum, M. fulvum*
3a	Macroconidia large (> 50 μm)		*Microsporum canis*
3b	Macroconidia smaller (< 50 μm)		*Microsporum equinum*
4a	Colony khaki-brown to greenish yellow		*Epidermophyton floccosum*
4b	Colony white, cream or clear yellow		*Trichophyton terrestre, T. ajelloi*
4c	Colony with some purple colour		*Trichophyton rubrum* granular form

Key to dermatophytes with microconidia predominant

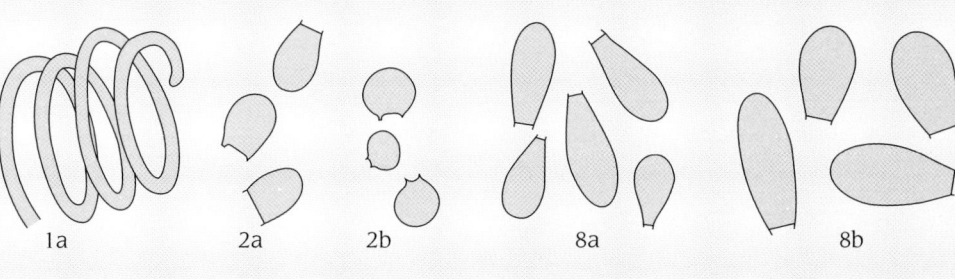

1a   2a   2b   8a   8b

1a	Spiral hyphae present	2
1b	Spiral hyphae absent	4
2a	Microconidia wide-based, often on ends of short branches	*Microsporum persicolor*
2b	Microconidia narrow-based, round	3
3a	Colony coarsely granular	*Trichophyton mentagrophytes*
3b	Colony powdery to suede	*Trichophyton interdigitale*
4a	Colony with yellow, glabrous fringed edge	5
4b	Colony without yellow, glabrous fringed edge	7
5a	Growth moderately fast (> 20 mm in one week)	6
5b	Growth slow (< 15 mm in one week)	*Trichophyton soudanense*
6a	Colony reverse dark brown	*Trichophyton equinum*
6b	Colony reverse deep orange	*Trichophyton interdigitale* nodular form
6c	Colony reverse bright yellow, diffusing pigment	*Trichophyton erinacei*
7a	Colony reverse bright yellow	*Trichophyton erinacei*
7b	Colony reverse cream, dark brown or red	8
8a	Microconidia small, round	*Trichophyton interdigitale*
8b	Microconidia small, club-shaped	9
8c	Microconidia larger, ovoid to club-shaped	10
9a	Reverse dark red brown with sharply defined white edge; urease negative at one week	*Trichophyton rubrum*
9b	Reverse white to cream, or if brown not sharply defined; urease positive at one week	*Trichophyton interdigitale* downy form
10a	Colony surface with some red or purple colour	*Trichophyton rubrum* granular form
10b	Colony surface with brown to yellow colours	*Trichophyton tonsurans*

Key to dermatophytes without macroconidia or microconidia

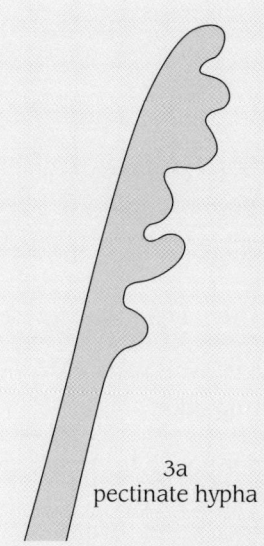
3a
pectinate hypha

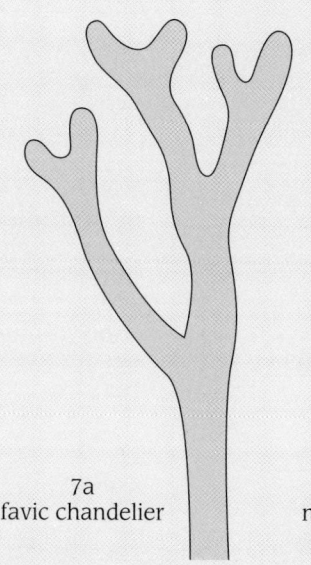
7a
favic chandelier

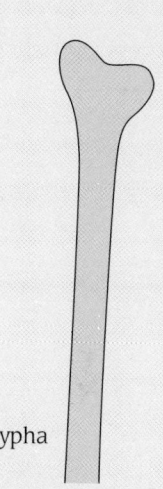
7a
nail-head hypha

1a	Growth moderately fast (> 15 mm in one week)	2
1b	Growth slow (< 15 mm in one week)	5
2a	Colony densely floccose	*Trichophyton rubrum*
2b	Colony loosely floccose or velvet	3
3a	Reverse pale pinkish-brown; pectinate hyphae may be present	*Microsporum audouinii*
3b	Reverse yellow	4
4a	Reverse intense orange-yellow	*Trichophyton interdigitale* nodular form
4b	Reverse clear yellow	*Microsporum canis*
5a	Colony dark purple-brown	*Trichophyton violaceum*
5b	Colony white, grey or cream	6
6a	Colony strongly folded	7
6b	Colony almost microscopic, floccose	*Trichophyton verrucosum*
7a	Nail-head hyphae and favic chandeliers present; grows without thiamine	*Trichophyton schoenleinii*
7b	Nail-head hyphae and favic chandeliers absent; most strains dependent on thiamine in medium	*Trichophyton concentricum*

MICROSPORUM CANIS

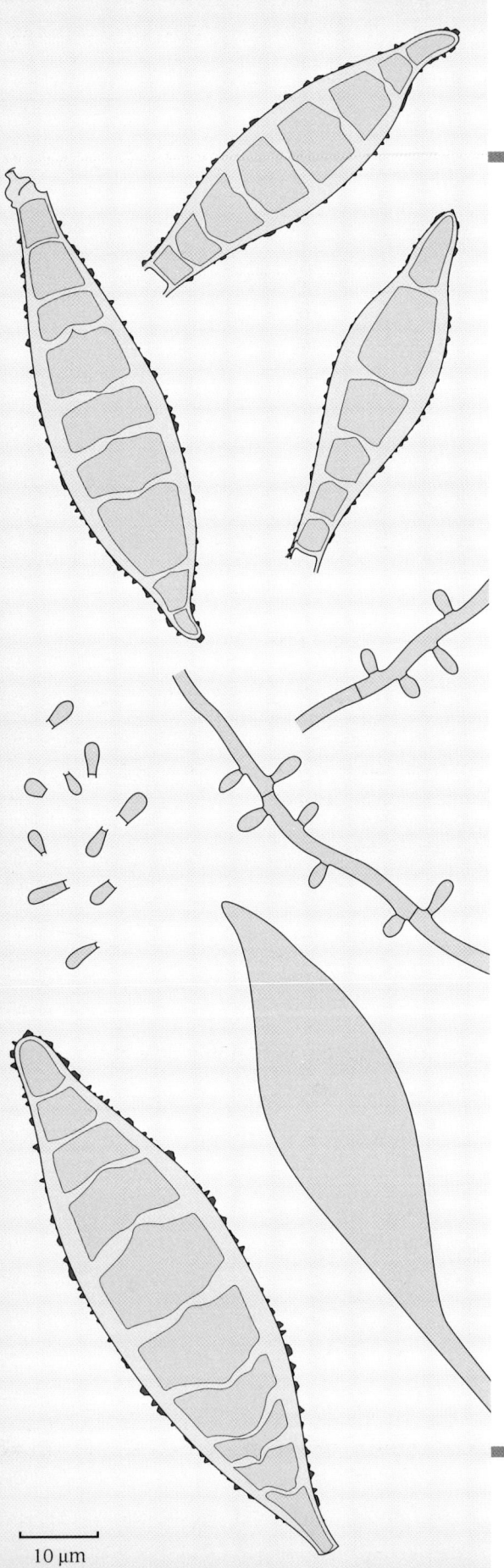

COLONIAL APPEARANCE
at 30°C on glucose peptone agar

diameter	50 mm in one week
topography	flat, with a few radial folds
texture	floccose, with wide, glabrous, radiating edge
colour	pale buff to white, with a yellow to colourless edge
reverse	bright yellow to colourless at the edge

MICROSCOPIC APPEARANCE
at 30°C

predominant features	large macroconidia, most common in the centre of the colonies
macroconidia	spindle-shaped, 35-110 µm × 12-25 µm; rough at least at the apical end, which is often curved to one side; most have 6-12 septa and very thick walls
microconidia	not common on glucose peptone agar, but more abundant on malt agar; narrow, club-shaped, borne along the sides of the hyphae

VARIANT FORMS

white form	lacks the yellow pigment, but otherwise identical
dysgonic (or glabrous) form	dense, low growth with irregular submerged edge and no spores; reflexive branching is common at the edge of colonies

DIFFERENTIAL DIAGNOSIS

colonial appearance	*Trichophyton erinacei*, which has a bright yellow pigment on the reverse and a pure white, granular top
	other *Microsporum* spp., which have a glabrous edge
	M. equinum resembles non-pigmented strains
- dysgonic form	*M. audouinii*, *T. soudanense*, the faviform *Trichophyton* spp. (*T. schoenleinii*, *T. verrucosum* and *T. concentricum*)
microscopic appearance	*M. equinum* and the rare dermatophyte *M. distortum* produce similar, but smaller macroconidia

SEXUAL STATE

Arthroderma otae

CLINICAL IMPORTANCE

It is a common cause of ectothrix tinea capitis and tinea corporis. The natural host is the domestic cat.

MICROSPORUM EQUINUM

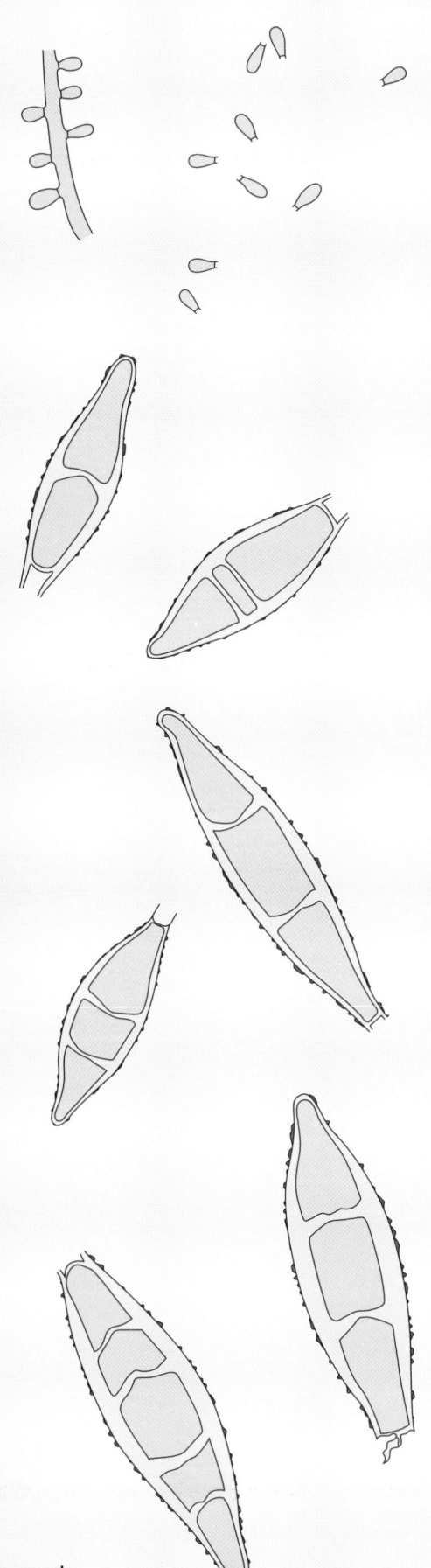

COLONIAL APPEARANCE
at 30°C on glucose peptone agar

diameter	40 mm in one week
topography	flat, with radial folds
texture	thinly floccose, with colourless, glabrous edge
colour	beige to cream
reverse	beige

MICROSCOPIC APPEARANCE
at 30°C

predominant features	occasional macroconidia, most common in the centre of the colonies
macroconidia	elliptical to spindle-shaped, 20-40 µm × 5-15 µm, with thick rough walls and two to three septa
microconidia	seldom found; pear- to club-shaped, borne along the sides of the hyphae

DIFFERENTIAL DIAGNOSIS

colonial appearance   *Microsporum audouinii, M. canis* strains without yellow pigmentation

microscopic appearance   *M. canis* produces similar but larger macroconidia with a more pronounced apical curvature

SEXUAL STATE

None known.

CLINICAL IMPORTANCE

It is an occasional cause of tinea corporis and ectothrix tinea capitis. The natural host is the horse.

MICROSPORUM GYPSEUM

M. gypseum

M. fulvum

10 µm

COLONIAL APPEARANCE
at 30°C on glucose peptone agar

diameter	40-50 mm in one week
topography	flat with radiating margin
texture	powdery
colour	buff to cinnamon, sometimes pink
reverse	buff to pinkish buff

MICROSCOPIC APPEARANCE
at 30°C

predominant features	numerous macroconidia, sparse microconidia
macroconidia	abundant, elliptical, 25-60 µm × 5-15 µm, with thin, roughened walls and four to six septa; sometimes with a terminal filament
microconidia	usually sparse, club-shaped, borne along the sides of the hyphae

DIFFERENTIAL DIAGNOSIS

colonial appearance *Microsporum fulvum*

microscopic appearance *M. fulvum*, but macroconidia of *M. gypseum* are broader

SEXUAL STATE

M. gypseum strains fall into two mating groups named *Arthroderma incurvatum* and *A. gypseum*.

CLINICAL IMPORTANCE

It is an occasional cause of tinea corporis and ectothrix tinea capitis. This geophilic dermatophyte has a worldwide distribution.

MICROSPORUM FULVUM

This is another geophilic dermatophyte that causes occasional human infection. On examination by the unaided eye or under the microscope it is very similar to *M. gypseum* but its macroconidia tend to be more slender (length to breadth ratio 4.5:1 compared with 3.5:1 for *M. gypseum*). Its sexual state is *A. fulvum*.

EPIDERMOPHYTON FLOCCOSUM

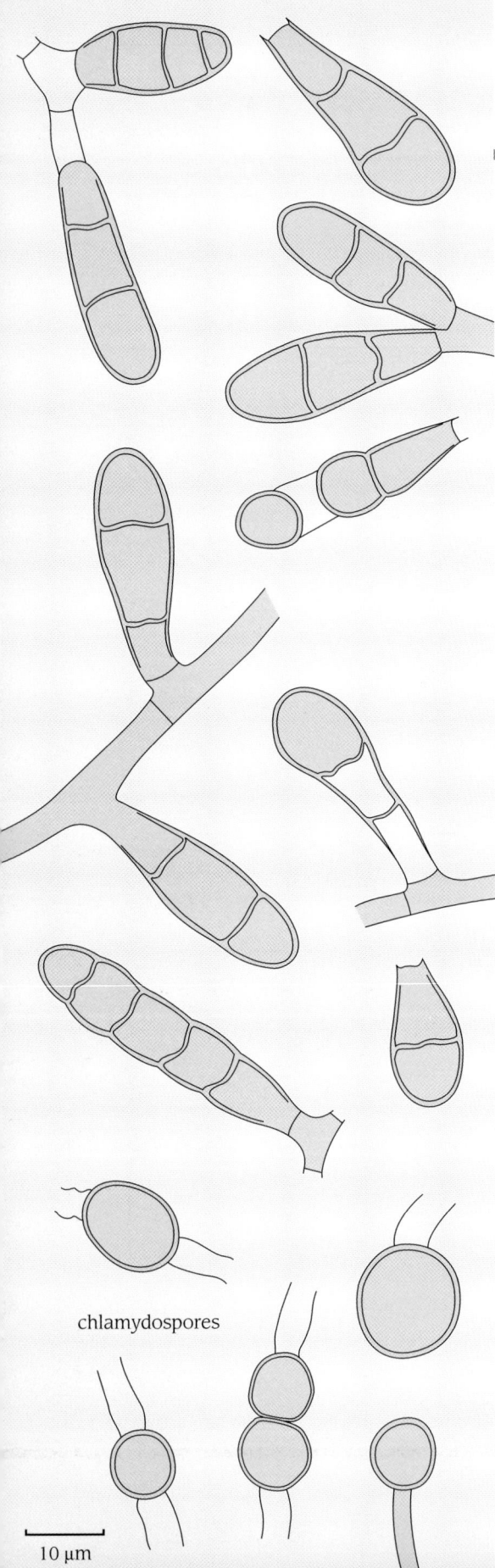

chlamydospores

10 μm

COLONIAL APPEARANCE
at 30°C on glucose peptone agar

diameter	15 mm in one week
topography	flat, sometimes folded in the centre
texture	powdery
colour	greenish yellow to khaki, with colourless submerged edge
reverse	pale brown

MICROSCOPIC APPEARANCE
at 30°C

predominant features	macroconidia, hyphal fragments and chlamydospores
macroconidia	oval to club-shaped, 25-45 μm × 8-10 μm, mostly with two to four septa; walls smooth and of medium thickness; often produced in clusters of two or three; soon lose cell contents
microconidia	absent

VARIANT FORMS

downy form	flat, floccose surface; fewer macroconidia, but these have up to six septa
pleomorphic form	white, floccose and without macroconidia; often develops within typical colonies

DIFFERENTIAL DIAGNOSIS

colonial appearance	*Trichophyton tonsurans*, *T. soudanense* and *Scopulariopsis brevicaulis* can produce similar yellowish-brown colonies
microscopic appearance	*T. tonsurans*, *T. violaceum* and other chlamydosporic dermatophytes resemble strains with abundant chlamydospores
	Chrysosporium keratinophilum produces similar-shaped but unicellular conidia

SEXUAL STATE

None known.

CLINICAL IMPORTANCE

It is a cause of tinea cruris worldwide and, less commonly, infection of the skin and nail in both hands and feet. It is confined to humans.

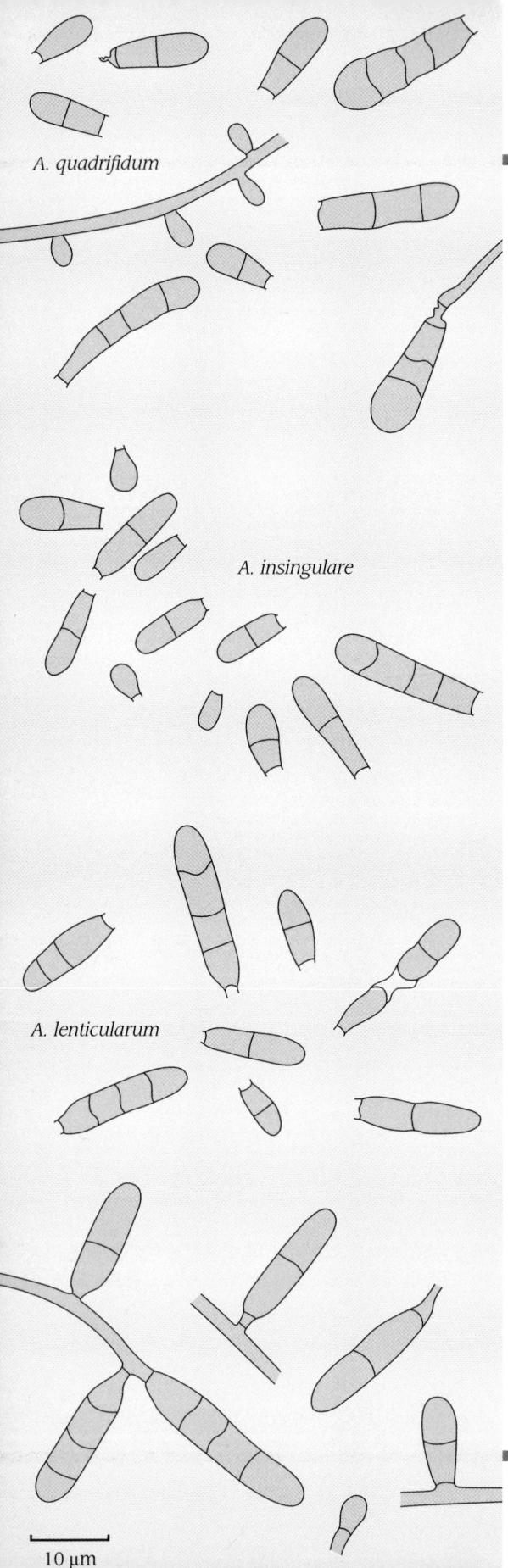

TRICHOPHYTON TERRESTRE

COLONIAL APPEARANCE
at 30°C on glucose peptone agar

diameter	10-30 mm in one week
topography	flat
texture	powdery to densely floccose
colour	white to pale cream
reverse	pale buff, yellow, to brown

MICROSCOPIC APPEARANCE
at 30°C

predominant features	macroconidia, microconidia and intermediate forms
conidia	most strains have a range of spores from oval, broad-based microconidia to cylindrical, thin-walled, broad-based macroconidia; two-celled conidia are common

DIFFERENTIAL DIAGNOSIS

colonial appearance *Trichophyton mentagrophytes*, *T. interdigitale* and other dermatophytes with flat, powdery colonies

microscopic appearance *T. mentagrophytes*, *T. ajelloi*

SEXUAL STATE

T. terrestre strains fall into three mating groups named *Arthroderma insingulare*, *A. lenticularum* and *A. quadrifidum*.

CLINICAL IMPORTANCE

It is normally not pathogenic but is an occasional contaminant of human skin and nail specimens.

MICROSPORUM PERSICOLOR

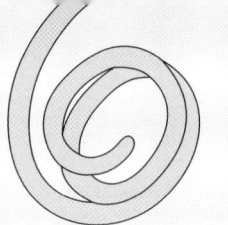
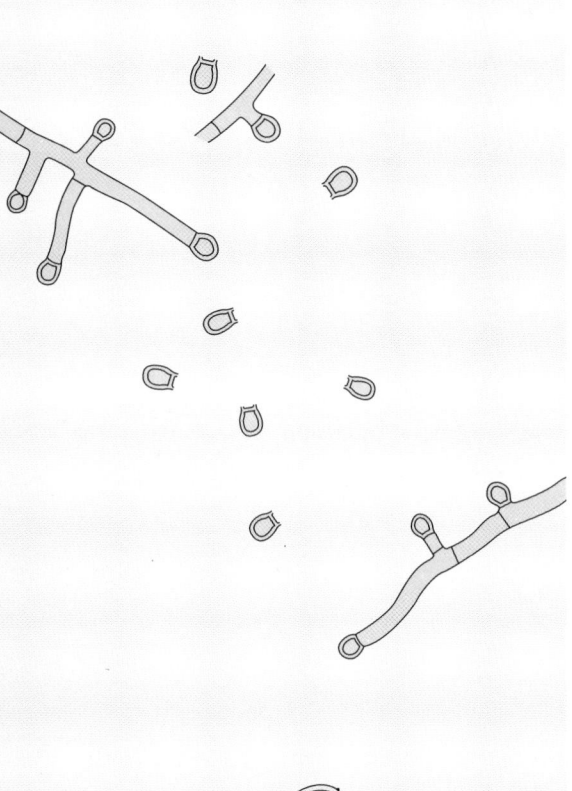
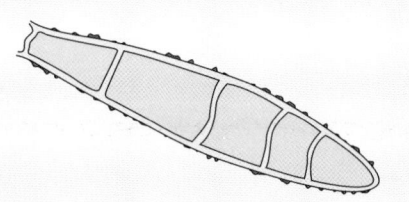

COLONIAL APPEARANCE
at 30°C on glucose peptone agar

diameter	20-25 mm in one week
topography	flat
texture	powdery to densely floccose
colour	cream to pinkish buff
reverse	light buff to brown

MICROSCOPIC APPEARANCE
at 30°C

predominant features	abundant microconidia, spiral hyphae and occasional macroconidia
macroconidia	narrow, cigar-shaped, 25-60 µm × 5-10 µm, mostly with three to five septa; walls thin, smooth to finely roughened
microconidia	abundant, round to oval, with a broad base; borne on short, lateral branches along the sides of the hyphae

DIFFERENTIAL DIAGNOSIS

colonial appearance *Trichophyton rubrum* and *T. mentagrophytes*, but *M. persicolor* often has a pinkish colour

microscopic appearance *T. mentagrophytes*, *T. interdigitale* and *T. equinum*, but *M. persicolor* macroconidia have finely roughened walls and its microconidia are distinctive

SEXUAL STATE

Arthroderma persicolor

CLINICAL IMPORTANCE

This geophilic dermatophyte has been isolated from small rodents, such as bank voles, and from dogs. It is an occasional cause of tinea corporis and tinea manuum.

TRICHOPHYTON EQUINUM

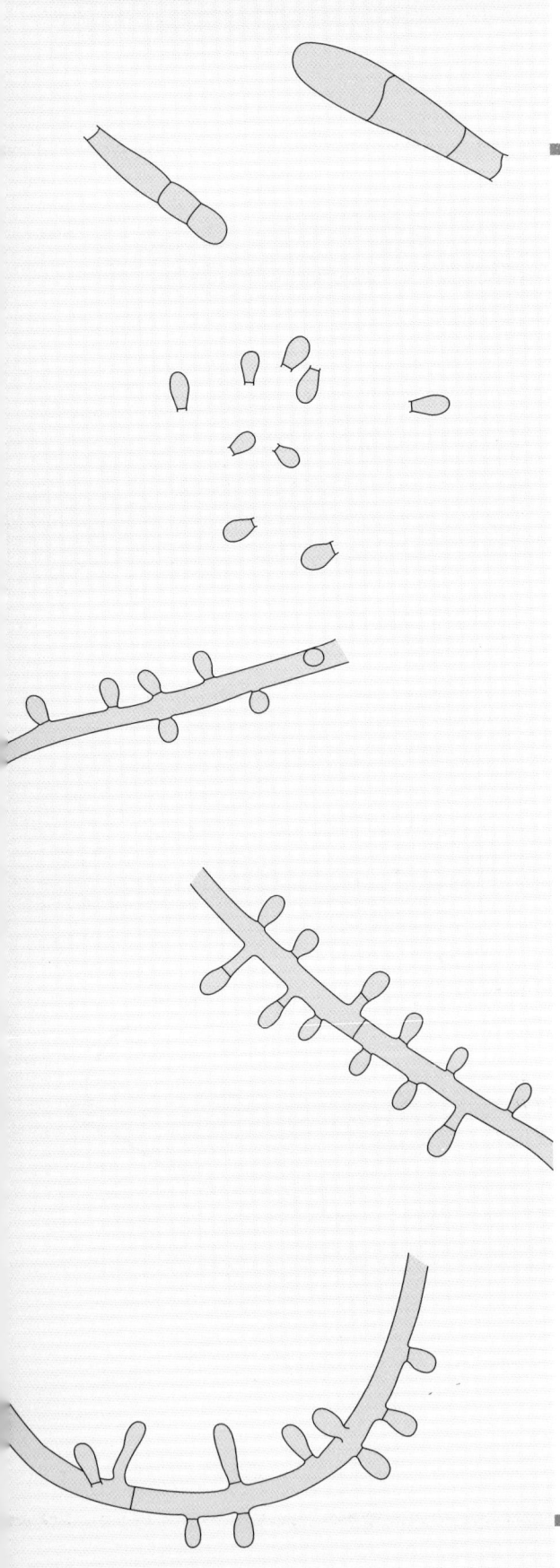

COLONIAL APPEARANCE
at 30°C on glucose peptone agar

diameter	30 mm in one week
topography	flat
texture	downy or densely floccose, often with a submerged edge
colour	white to cream
reverse	yellow, especially at the edge; centre becomes reddish brown to dark brown with age; some strains produce a diffusing, brown pigment

MICROSCOPIC APPEARANCE
at 30°C

predominant features	abundant microconidia
macroconidia	seldom found; club-shaped, 10-65 µm × 4-12 µm, with thin, smooth walls and two to three septa
microconidia	abundant; club-shaped, borne along the sides of the hyphae

10 µm

VARIANT FORM

var. *autotrophicum*   does not require nicotinic acid in the growth medium

DIFFERENTIAL DIAGNOSIS

colonial appearance   *Trichophyton interdigitale, T. mentagrophytes, T. erinacei* and *T. rubrum* all share some features with *T. equinum*, but the latter requires nicotinic acid in the growth medium

microscopic appearance   *T. mentagrophytes, T. erinacei, T. interdigitale, Microsporum persicolor*

SEXUAL STATE

None known.

CLINICAL IMPORTANCE

It is an occasional cause of tinea corporis and ectothrix tinea capitis. The natural host is the horse.

TRICHOPHYTON ERINACEI

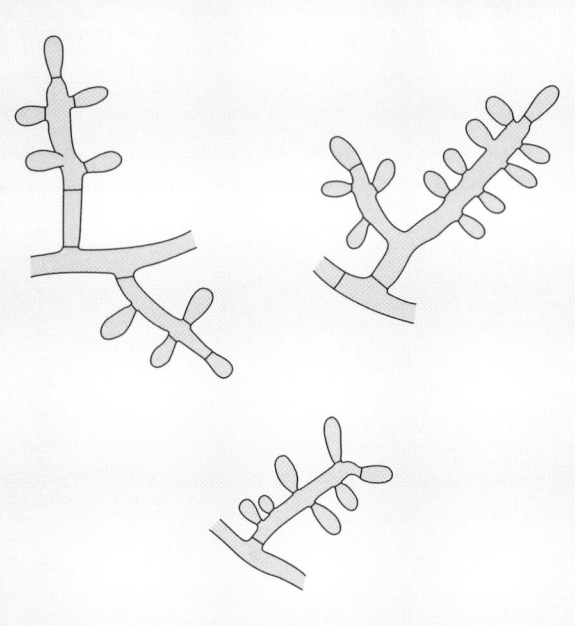

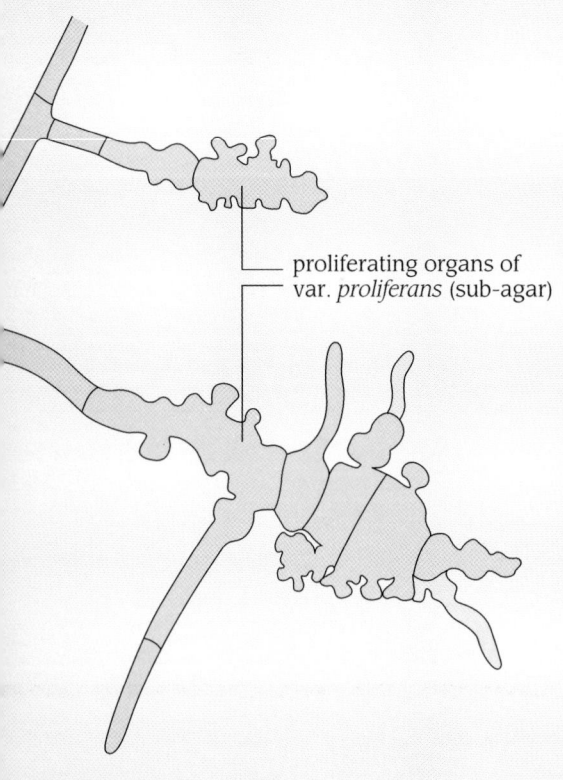

proliferating organs of var. *proliferans* (sub-agar)

10 μm

COLONIAL APPEARANCE
at 30°C on glucose peptone agar at 30°C

diameter	20-30 mm in one week
topography	flat
texture	granular to powdery
colour	white
reverse	bright yellow, diffusing into medium

MICROSCOPIC APPEARANCE
at 30°C

predominant features	abundant microconidia
macroconidia	sparse; cylindrical, with thin, smooth walls and three to five septa
microconidia	club-shaped, borne along the sides and ends of branched hyphae

VARIANT FORM

var. *proliferans*   powdery to floccose colonies with a buff to orange reverse; few microconidia; characteristic hyphal propagules

DIFFERENTIAL DIAGNOSIS

colonial appearance   *Microsporum canis*, yellow form of *Trichophyton rubrum*, nodular form of *T. interdigitale* and other dermatophytes that form yellow colonies

microscopic appearance   *T. rubrum*, downy form of *T. interdigitale*, *T. equinum*, *T. mentagrophytes* strains with club-shaped microconidia

Note: *T. erinacei* strains from British sources are urease negative, but *M. canis* and *T. interdigitale* are mostly urease positive.

SEXUAL STATE

Arthroderma benhamiae

CLINICAL IMPORTANCE

It is an uncommon cause of tinea corporis. The European hedgehog is the natural host, but infection of dogs is not uncommon.

TRICHOPHYTON MENTAGROPHYTES

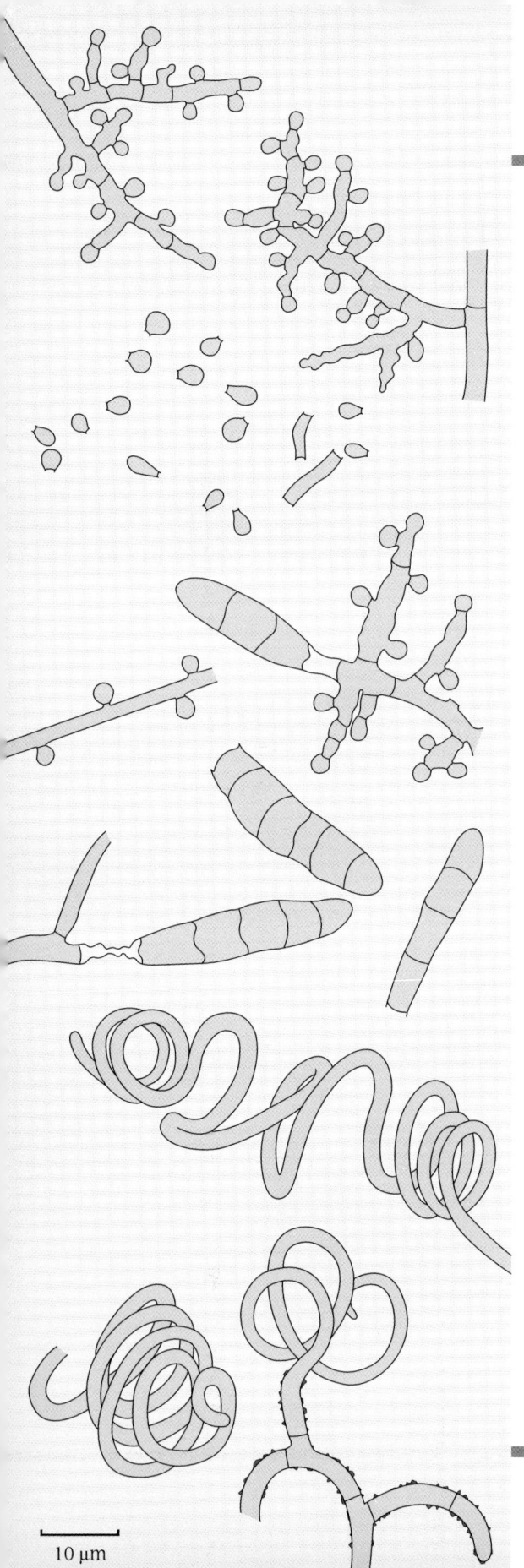

COLONIAL APPEARANCE
at 30°C on glucose peptone agar at 30°C

diameter	20-30 mm in one week
topography	flat
texture	granular to powdery
colour	white to cream, sometimes pale pink, grey or yellow
reverse	cream to dark brown, often with radiating brown striations

MICROSCOPIC APPEARANCE
at 30°C

predominant features	abundant microconidia, some macroconidia, and spiral hyphae
macroconidia	quite common; cylindrical, 20-50 µm × 7-10 µm, with thin, smooth walls and mostly three to four septa
microconidia	predominantly round, borne along the sides and ends of repeatedly branched hyphae to form large clusters

10 µm

VARIANT FORMS

Forms related to one sexual state (*Arthroderma benhamiae*) have club-shaped microconidia resembling *Trichophyton erinacei*.

DIFFERENTIAL DIAGNOSIS

colonial appearance	powdery forms of *T. interdigitale* and *T. erinacei*, granular forms of *T. rubrum*, *T. tonsurans*, *T. terrestre*, some *Chrysosporium* spp.
microscopic appearance	*T. interdigitale*, *T. terrestre*
	T. erinacei, *T. tonsurans* and *T. equinum* resemble forms with club-shaped microconidia

SEXUAL STATE

T. mentagrophytes strains fall into two mating groups named *Arthroderma benhamiae* and *A. vanbreuseghemii*. Most European strains are compatible with *A. vanbreuseghemii*.

CLINICAL IMPORTANCE

It is an uncommon cause of tinea corporis and ectothrix tinea capitis and widespread among animals, particularly rodents.

TRICHOPHYTON INTERDIGITALE

downy form

10 µm

COLONIAL APPEARANCE
at 30°C on glucose peptone agar

diameter	20-30 mm in one week
topography	flat, sometimes folded
texture	powdery to suede
colour	white with cream centre, sometimes pink or grey
reverse	cream to dark brown

MICROSCOPIC APPEARANCE
at 30°C

predominant features	abundant microconidia, occasional macroconidia and spiral hyphae
macroconidia	sparse; cylindrical with thin, smooth walls and three to four septa
microconidia	predominantly round, borne along the sides and ends of repeatedly branched hyphae

VARIANT FORMS

downy form	high, white, floccose colonies with a buff reverse; microconidia club-shaped to round; elongated arthrospores present
nodular form	bright orange-yellow colonies with a little aerial mycelium, usually as a central white tuft; 'nodular bodies' or hyphal knots may be seen; sparse, round microconidia

DIFFERENTIAL DIAGNOSIS

colonial appearance	*Trichophyton mentagrophytes*, *T. erinacei*, *T. equinum*, *T. terrestre*, granular form of *T. rubrum*, some *Chrysosporium* spp.
- downy form	poorly pigmented strains of *T. rubrum*, but these are urease negative
microscopic appearance	*T. mentagrophytes* is indistinguishable from the more powdery forms
	T. rubrum and *T. erinacei* are similar to the downy form but their microconidia are less rounded

SEXUAL STATE

Some *T. interdigitale* strains are compatible with *Arthroderma vanbreuseghemii* or *A. benhamiae* but most are incompatible.

CLINICAL IMPORTANCE

It is a common cause of tinea pedis and tinea unguium and is confined to humans.

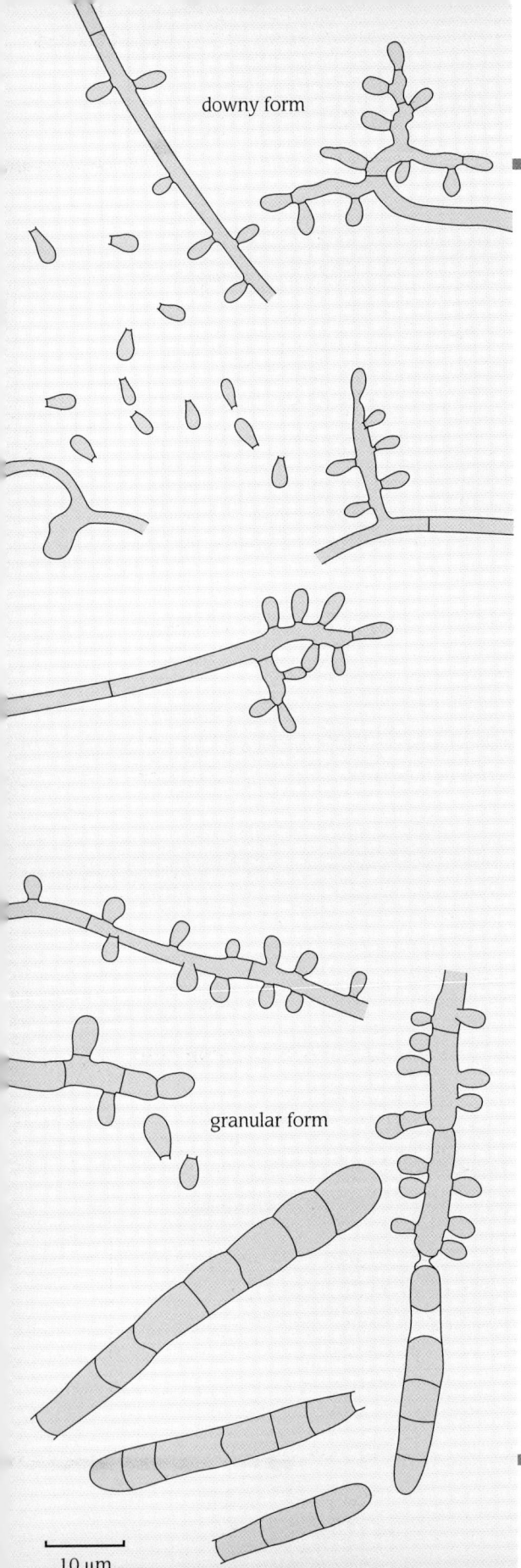

TRICHOPHYTON RUBRUM

COLONIAL APPEARANCE
at 30°C on glucose peptone agar

diameter	10-15 mm in one week
topography	domed
texture	downy to floccose
colour	white
reverse	dark red-brown with a sharply demarcated white edge

MICROSCOPIC APPEARANCE
at 30°C

predominant features	sparse microconidia; sometimes a few arthrospores
macroconidia	absent
microconidia	club-shaped; formed along the sides of the hyphae

52

VARIANT FORMS

granular form	see below
melanin-producing form	diffusing, dark-brown pigment
yellow form	no spores, pale yellow aerial growth and yellow reverse
glabrous form	dark-red wrinkled colonies, rare

DIFFERENTIAL DIAGNOSIS

colonial appearance	*Trichophyton interdigitale*, except that *T. rubrum* colonies have a well-defined zone of pigmentation on the reverse
- glabrous form	*T. violaceum*, except that *T. rubrum* usually produces some microconidia
- yellow form	*T. erinacei*
microscopic appearance	*T. interdigitale*

Note: The urease test distinguishes between *T. interdigitale* and *T. rubrum* except for the granular form of *T. rubrum*.

SEXUAL STATE

None known.

CLINICAL IMPORTANCE

It is a common cause of tinea pedis, tinea cruris, tinea corporis, tinea manuum and tinea unguium. It is confined to humans.

TRICHOPHYTON RUBRUM (GRANULAR FORM)

This type is common in East Asia. Colonies are flat to folded, granular, cream to purple with a dark red-brown reverse. Microconidia are larger than in the typical downy type. Macroconidia are cylindrical, 20-50 µm × 4-6 µm, with up to eight septa. Arthrospores are frequently present. The colonial appearance can be confused with *T. mentagrophytes* and *T. erinacei* and the microscopic appearance with *T. tonsurans* and *T. soudanense*.

TRICHOPHYTON TONSURANS

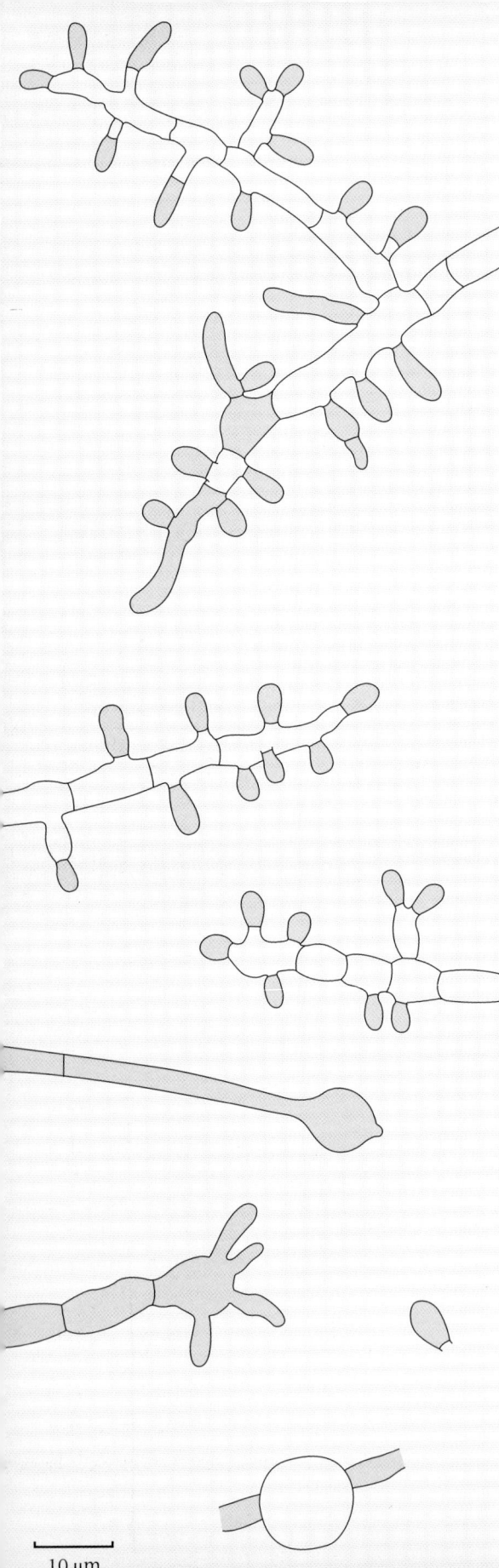

COLONIAL APPEARANCE
at 30°C on glucose peptone agar

diameter	15-20 mm in one week
topography	flat, sometimes folded in centre, occasionally with surface cracks
texture	powdery to suede-like
colour	cream-white to pale brown
reverse	yellow to reddish brown

MICROSCOPIC APPEARANCE
at 30°C

predominant features	abundant microconidia, chlamydospores
macroconidia	sparse; cylindrical, 10-50 µm × 4-8 µm, with up to ten septa; often become distorted with age
microconidia	oval to club-shaped, large, borne along the sides of widened hyphae, which often lose their contents; individual microconidia may become greatly enlarged ('balloon forms')

VARIANT FORM

var. *sulphureum*   yellow aerial growth and a yellow reverse

DIFFERENTIAL DIAGNOSIS

colonial appearance   *Trichophyton mentagrophytes*, *T. soudanense*, granular form of *T. rubrum*, *Epidermophyton floccosum*

microscopic appearance   *T. mentagrophytes*, *T. soudanense*, granular form of *T. rubrum*, chlamydosporic strains of *E. floccosum*

SEXUAL STATE

None known.

CLINICAL IMPORTANCE

It is a cause of tinea corporis, endothrix tinea capitis and, rarely, infection of other sites. It is worldwide in distribution but commonest in the Americas. It is confined to humans.

TRICHOPHYTON SOUDANENSE

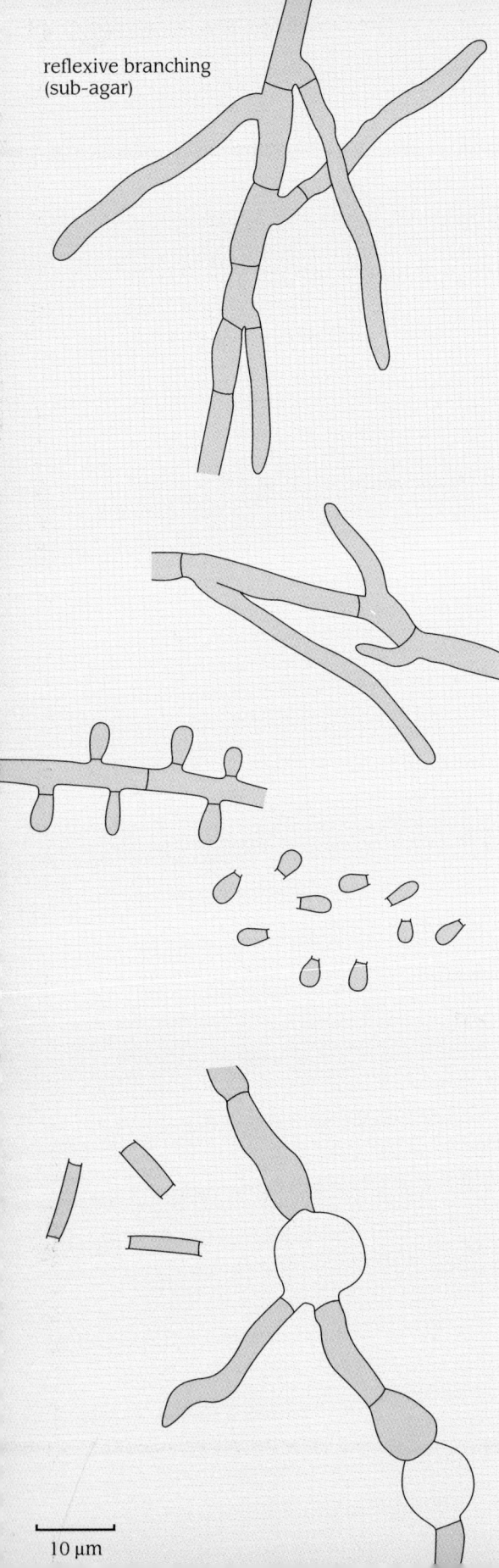

reflexive branching (sub-agar)

10 μm

COLONIAL APPEARANCE
at 30°C on glucose peptone agar

diameter	10 mm in one week
topography	flat, with heaped, folded centre
texture	glabrous, with submerged edge
colour	deep orange-yellow, often with a reddish centre
reverse	deep orange-yellow

MICROSCOPIC APPEARANCE
at 30°C

predominant features	few microconidia; reflexive branching at edge of colonies; arthrospores and chlamydospores may occur
macroconidia	absent
microconidia	large, oval, borne along the sides of the hyphae

DIFFERENTIAL DIAGNOSIS

colonial appearance   *Epidermophyton floccosum*, *Trichophyton tonsurans*, nodular form of *T. interdigitale*, dysgonic form of *Microsporum canis*; *T. violaceum* resembles reddish forms

microscopic appearance   *T. tonsurans*, granular form of *T. rubrum*

SEXUAL STATE

None known.

CLINICAL IMPORTANCE

It is a cause of tinea corporis and endothrix tinea capitis and is commonest in Africa.

MICROSPORUM AUDOUINII

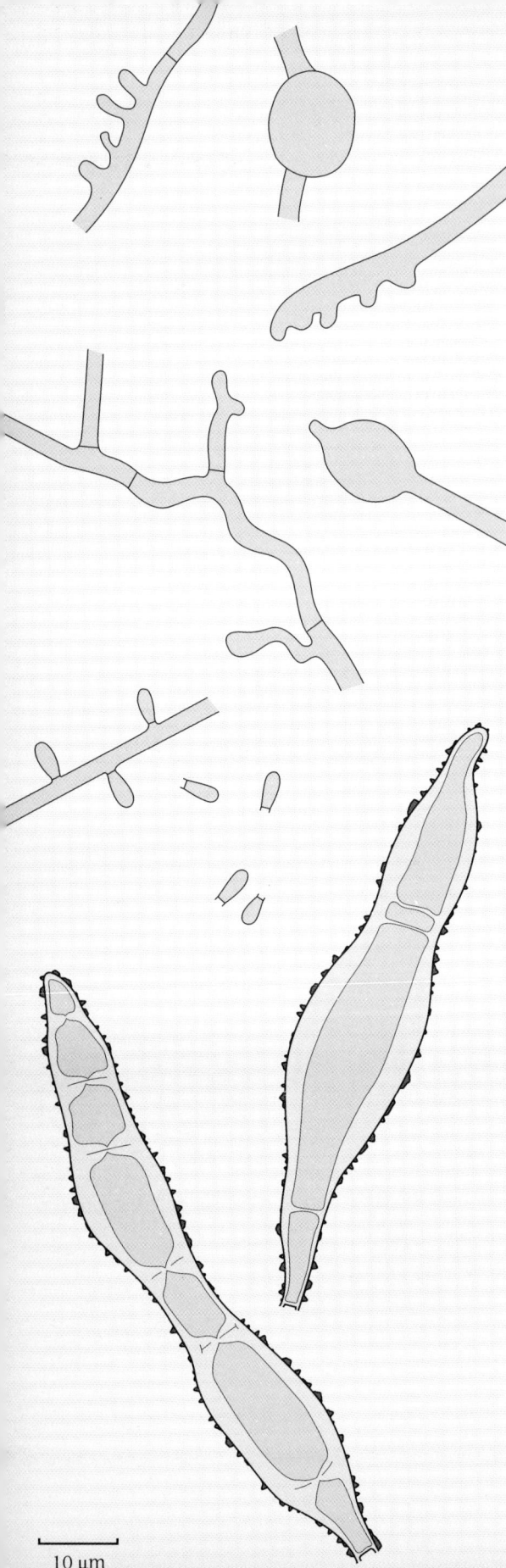

COLONIAL APPEARANCE
at 30°C on glucose peptone agar

diameter	20 mm in one week
topography	flat
texture	thinly floccose, with glabrous edge
colour	pale buff, sometimes with a whitish aerial mycelium
reverse	deep apricot brown

MICROSCOPIC APPEARANCE
at 30°C

predominant features	hyphal growth; pectinate hyphae sometimes present
macroconidia	seldom found; spindle-shaped, 30-80 µm × 8-14 µm, with thick rough walls and one to four septa
microconidia	seldom found; club-shaped, 3-9 µm × 1.5-3 µm, borne along the sides of the hyphae

VARIANT FORMS

var. *langeronii*   has a more pronounced brown colour

var. *rivalieri*   has a grey, much folded surface, with a flat margin

DIFFERENTIAL DIAGNOSIS

colonial appearance   dysgonic form of *Microsporum canis*, *M. equinum*

microscopic appearance   any non-sporing floccose mould

SEXUAL STATE

None known.

CLINICAL IMPORTANCE

It is a significant cause of ectothrix tinea capitis in children in Central and West Africa and the Americas. It is confined to humans.

TRICHOPHYTON SCHOENLEINII

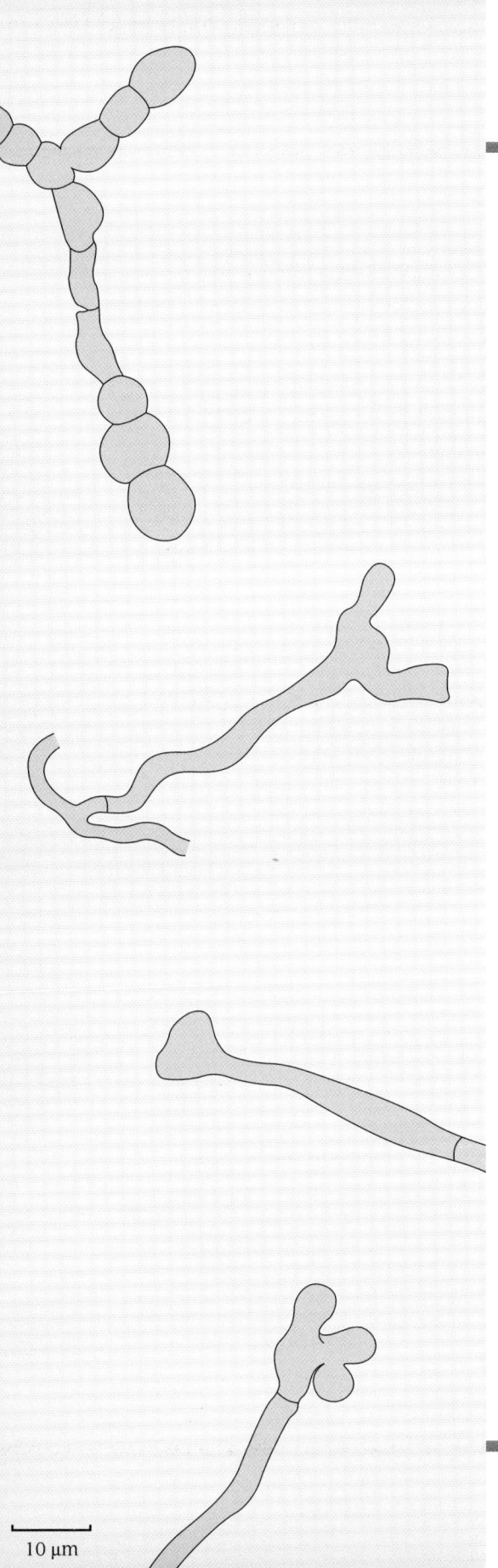

COLONIAL APPEARANCE
at 30°C on glucose peptone agar

diameter	5-10 mm in one week
topography	heaped, irregularly folded, with a submerged edge
texture	glabrous, with little aerial mycelium
colour	white to light grey or buff
reverse	colourless

MICROSCOPIC APPEARANCE
at 30°C

predominant features	characteristic antler hyphae (favic chandeliers) with swollen 'nail head' tips
macroconidia	absent
microconidia	absent

10 μm

DIFFERENTIAL DIAGNOSIS

colonial appearance
Trichophyton verrucosum, dysgonic form of *Microsporum canis*; some non-dermatophytes (on media containing cycloheximide)

microscopic appearance
T. verrucosum and *T. violaceum* both produce distorted hyphae, but the characteristic 'nail head' hyphae are infrequent

SEXUAL STATE

None known.

CLINICAL IMPORTANCE

It is the cause of favus in humans, a condition commonest in North Africa, the Middle East and parts of southern and eastern Europe.

TRICHOPHYTON VERRUCOSUM

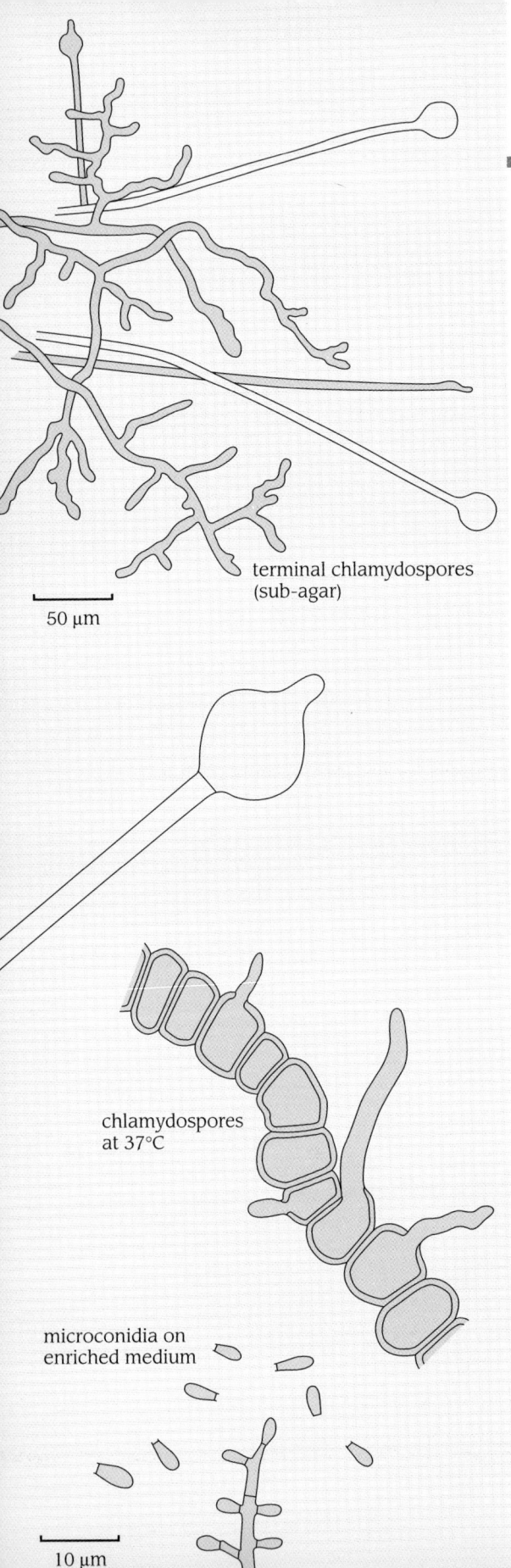

terminal chlamydospores (sub-agar)

50 μm

chlamydospores at 37°C

microconidia on enriched medium

10 μm

COLONIAL APPEARANCE
at 30°C on glucose peptone agar

diameter	less than 5 mm in one week
topography	flat, sometimes domed
texture	glabrous or floccose
colour	white to cream
reverse	white to cream

MICROSCOPIC APPEARANCE
at 30°C

predominant features	hyphal growth; characteristic large, empty terminal chlamydospores on the submerged mycelium; at 37°C thick-walled chlamydospores are produced in chains
macroconidia	not formed on normal media
microconidia	club-shaped microconidia formed on enriched media

VARIANT FORM

var. *ochraceum*   flat, yellow colonies

DIFFERENTIAL DIAGNOSIS

colonial appearance   most strains are very slow growing and unlikely to be confused with other dermatophytes

microscopic appearance   *Trichophyton rubrum* may resemble floccose strains with microconidia

SEXUAL STATE

None known.

CLINICAL IMPORTANCE

It is a cause of tinea corporis and ectothrix tinea capitis. Kerion formation is common. Cattle are the natural host.

TRICHOPHYTON VIOLACEUM

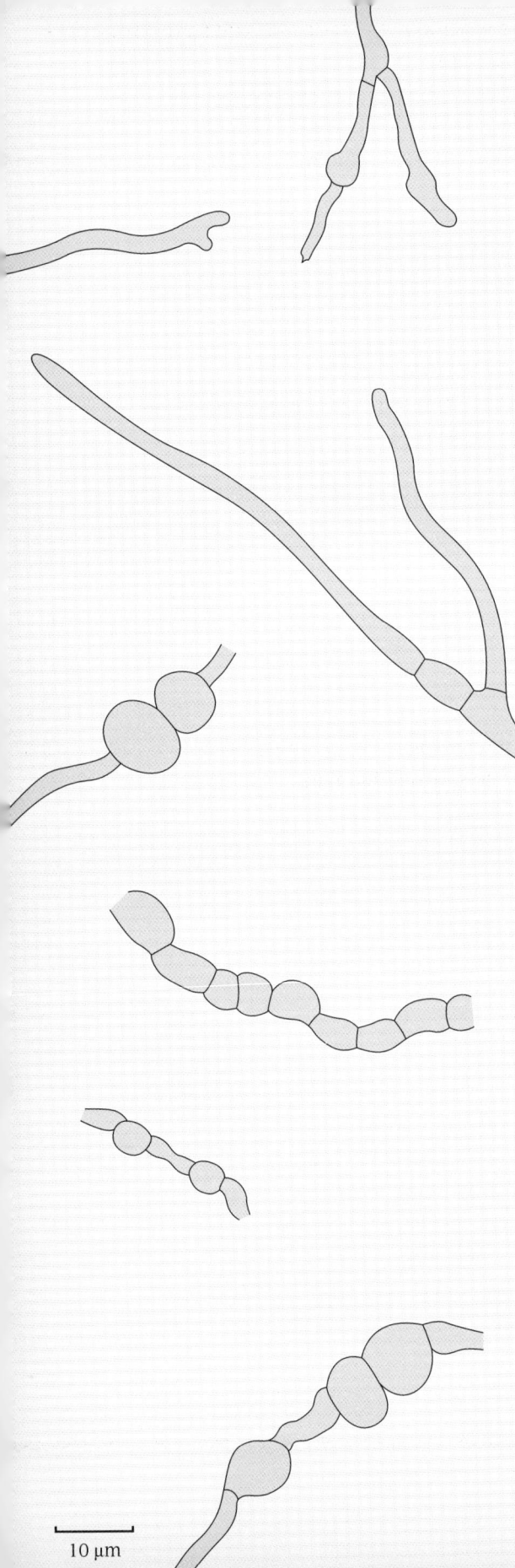

COLONIAL APPEARANCE
at 30°C on glucose peptone agar

diameter	10 mm in one week
topography	irregularly wrinkled; heaped in the centre
texture	glabrous
colour	dark purple-red
reverse	dark purple-red

MICROSCOPIC APPEARANCE
at 30°C

predominant features	hyphal growth; chlamydospores sometimes present
macroconidia	absent
microconidia	pear-shaped; only seen on enriched media

VARIANT FORM

pale form   lacks the purple-red pigment

DIFFERENTIAL DIAGNOSIS

colonial appearance   *Trichophyton gourvilii,* rare glabrous form of *T. rubrum*

- pale form   *T. yaoundei*

microscopic appearance   *T. concentricum, T. schoenleinii, T. verrucosum*

SEXUAL STATE

None known.

CLINICAL IMPORTANCE

It is one of the commonest causes of endothrix tinea capitis in eastern Europe, Asia, North Africa, and Central and South America. It is confined to humans.

TRICHOPHYTON CONCENTRICUM

COLONIAL APPEARANCE
at 30°C on glucose peptone agar

diameter	10 mm in one week
topography	heaped; highly folded
texture	glabrous, covered with short hyphae
colour	white, yellow-brown, orange-brown
reverse	yellow-brown

MICROSCOPIC APPEARANCE
at 30°C

predominant features	tangled, branching hyphae; no microconidia or macroconidia; chlamydospores may be present

colonial topography

10 μm

VARIANT FORM

indicum form   flat, slightly downy colonies

DIFFERENTIAL DIAGNOSIS

colonial appearance   *Trichophyton schoenleinii*, but that is usually more glabrous and deeper brown

microscopic appearance   *T. schoenleinii*, except that the terminal hyphae of *T. concentricum* lack the swollen nail-head ends

SEXUAL STATE

None known.

CLINICAL IMPORTANCE

It is the cause of tinea imbricata in native populations of south Pacific islands, South East Asia, and parts of Central and South America. *T. concentricum* does not invade hair and this may help in distinguishing it from *T. schoenleinii* which causes classical favus.

OTHER *MICROSPORUM* AND *TRICHOPHYTO...*

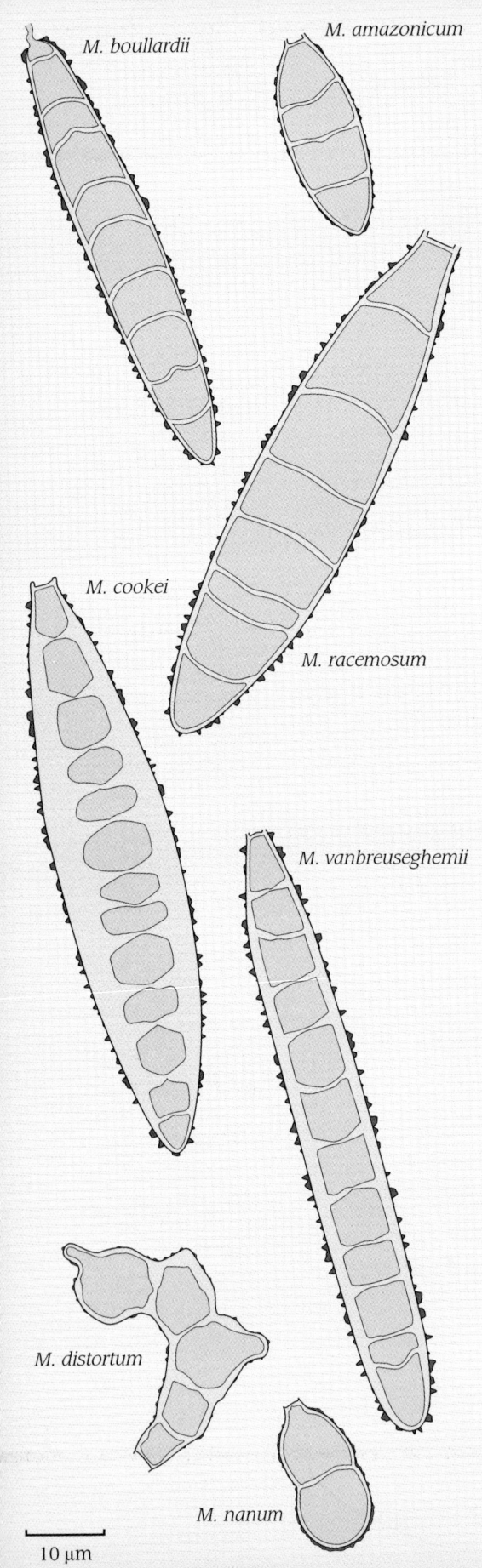

MICROSPORUM BOULLARDII

colony   powdery, pink to buff surface

microscopy   abundant, thin-walled, roughened macroconidia; abundant microconidia

clinical importance   soil organism

MICROSPORUM AMAZONICUM

colony   flat, velvety, grey-brown surface

microscopy   abundant, small, oval, roughened macroconidia; microconidia also present

clinical importance   soil organism

MICROSPORUM RACEMOSUM

colony   powdery, flat, cream-coloured surface; reverse deep wine-red

microscopy   abundant, thin-walled, roughened macroconidia; abundant microconidia

clinical importance   soil organism; very rare cause of human infection

MICROSPORUM COOKEI

colony	coarsely powdery to granular, cream to pink-buff; reverse deep red
microscopy	thick-walled, roughened macroconidia; abundant microconidia
clinical importance	soil organism, reported in small mammals; very rare cause of human infection

MICROSPORUM VANBREUSEGHEMII

colony	fast-growing, flat, powdery to floccose, cream to pink-buff
microscopy	abundant, thick-walled, roughened macroconidia; abundant microconidia
clinical importance	soil organism, reported in small mammals; occasional cause of human infection; ectothrix hair invasion

MICROSPORUM DISTORTUM

colony	loosely floccose, white to pale buff
microscopy	abundant microconidia; scarce, distorted macroconidia
clinical importance	rare cause of human infection; ectothrix hair invasion

MICROSPORUM NANUM

colony	powdery, cream to buff surface; reverse dull red-brown
microscopy	abundant, two-celled, rough-walled macroconidia; scarce microconidia
clinical importance	probable soil organism often causing infection in pigs; rare cause of human infection; ectothrix hair invasion

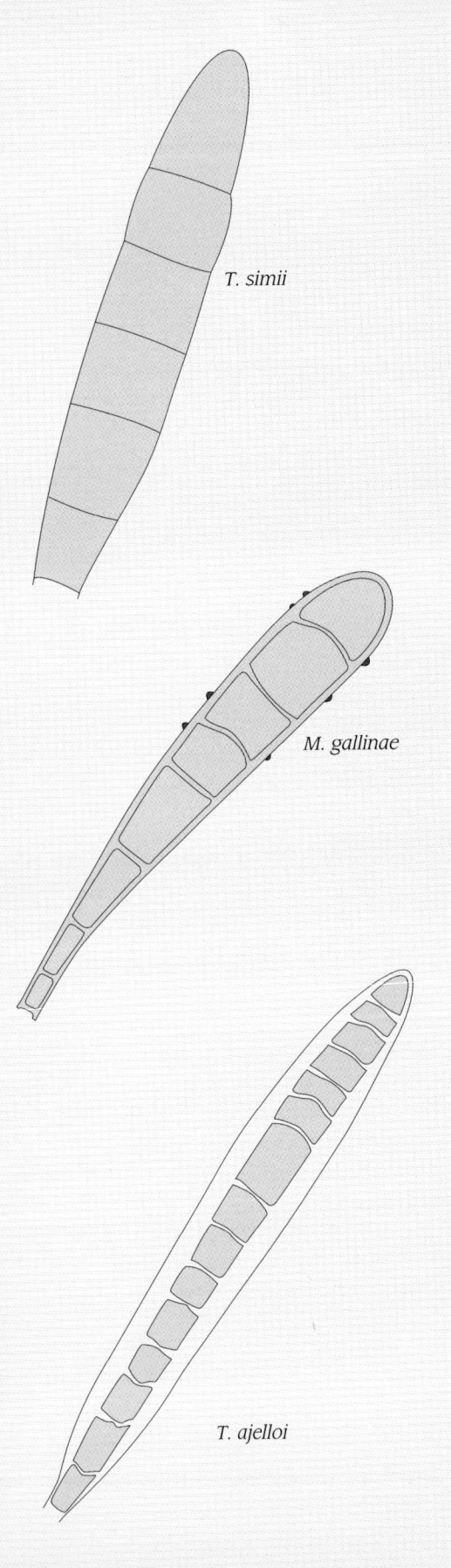

MICROSPORUM GALLINAE

colony	downy to velvety, white, turning pink with age; reverse orange-pink with diffusing, red pigment
microscopy	club-shaped macroconidia, up to ten septa, often bent and tapered towards the base; abundant microconidia
clinical importance	infects birds; a rare cause of human infection

TRICHOPHYTON SIMII

colony	rapid growing, granular to powdery, white to buff; reverse yellow to reddish brown in centre
microscopy	abundant macroconidia; intercalary chlamydospores sometimes formed within the macroconidia; abundant microconidia
clinical importance	soil organism; infects monkeys and chickens in India; occasional human infection

TRICHOPHYTON AJELLOI

colony	powdery, orange to buff; reverse deep purple to black
microscopy	abundant macroconidia; microconidia usually absent
clinical importance	soil organism; doubtful cause of human infection

TRICHOPHYTON GOURVILII

colony	folded and heaped, glabrous, pink to red; sometimes with diffusing, brown pigment
microscopy	occasional long, club-shaped, thin-walled macroconidia; abundant microconidia
clinical importance	cause of tinea capitis in West Africa; endothrix hair invasion

TRICHOPHYTON MEGNINII

colony	velvety with radial grooves, pale pink; reverse pale red at centre
microscopy	abundant microconidia; rare, narrow, club-shaped macroconidia
clinical importance	rare cause of human infection in Africa and Europe; ectothrix hair invasion

TRICHOPHYTON YAOUNDEI

colony	slow growing, glabrous grey, submerged or heaped and folded, becoming brown with age; diffusing, brown pigment
microscopy	usually hyphae only; rarely microconidia
clinical importance	cause of tinea capitis in equatorial Africa; endothrix hair invasion

5 MOULDS WITH ALEURIOSPORES I

INTRODUCTION

Like the dermatophytes, most other aleuriosporic moulds are related to a single fungal group, the Order *Onygenales*. Many are soil saprophytes with strong keratinolytic or proteolytic activity and some are potentially highly infectious (e.g. *Histoplasma capsulatum, Blastomyces dermatitidis* and *Paracoccidioides brasiliensis* are classified as Hazard Group 3 pathogens). *Myceliophthora thermophila* and *Chrysosporium keratinophilum* are occasional causes of deep and superficial infections respectively. These and other species of the *Chrysosporium-Myceliophthora* group may be encountered as laboratory contaminants.

All the fungi described in this chapter grow without inhibition on media containing cycloheximide, and many have white or cream, floccose or powdery colonies. In addition they give a red colour reaction on dermatophyte test medium. These species do, however, show features that may serve to distinguish them from the dermatophytes. Apart from *Geomyces pannorum*, the aleuriospores are either much larger than dermatophyte microconidia or round with inconspicuous scars. *Chrysosporium* species mostly have large, club-shaped spores borne on the sides or ends of normal hyphal cells. The spores in some *Myceliophthora* species may also be club-shaped but differ from *Chrysosporium* species in that at least some arise from swollen hyphal cells. Other *Myceliophthora* species have large, round conidia resembling Histoplasma macroconidia.

The identification of *H. capsulatum, B. dermatitidis* and *P. brasiliensis* should always involve demonstration of specific exoantigens and/or conversion to the yeast phase. Conversion of the mycelial phase to the yeast phase requires special media such as blood-glucose-cysteine agar and incubation at 37°C for several weeks. The size, shape and manner of budding of these yeasts are characteristic for each species and form useful identification features.

Key to aleuriosporic moulds other than dermatophytes

1a	Conidia absent (yeast phase of large globose cells with multiple budding)	*Paracoccidioides brasiliensis*
1b	Conidia present	2
2a	Conidia large, spherical, rough walled or tuberculate	*Histoplasma capsulatum*
2b	Conidia not as above	3
3a	Conidia small (< 4 μm long)	4
3b	Conidia larger	5
4a	Conidia broad-based on acutely branched conidiophores	*Geomyces pannorum*
4b	Conidia with very narrow scars, on sides of hyphae or on short stalks	*Blastomyces dermatitidis*
5a	Colony cinnamon-brown in centre, some conidia on swollen hyphal cells	*Myceliophthora thermophila*
5b	Colony white to cream in centre, conidia not on swollen hyphal cells	*Chrysosporium keratinophilum*

GEOMYCES PANNORUM

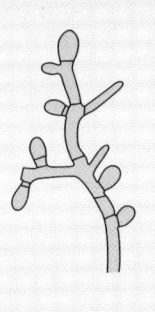
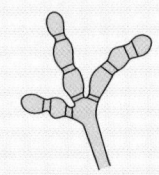
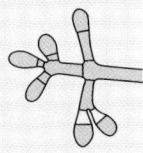
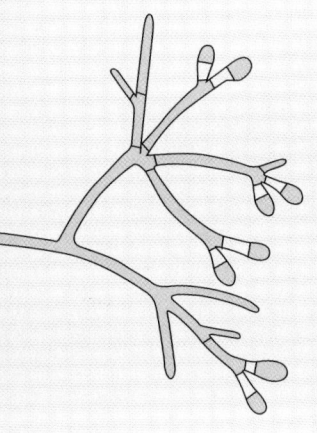
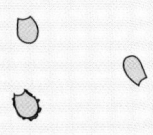
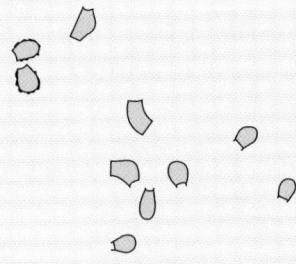

COLONIAL APPEARANCE
at 30°C on glucose peptone agar

diameter	10 mm in one week
topography	flat, sometimes heaped
texture	granular to powdery
colour	cream to grey-buff
reverse	pale buff

MICROSCOPIC APPEARANCE
at 30°C

predominant features	abundant, small conidia
conidia	oval, with flat base where attached to conidiophore; terminal or lateral on short, acutely branched conidiophores; arthrospores produced in alternate segments of the conidiophore

10 µm

VARIANT FORMS

var. *asperulatus*	yellow colonies with chains of conidia
var. *vinaceus*	reddish-brown colonies with diffusing, red pigment

DIFFERENTIAL DIAGNOSIS

colonial appearance	slow-growing dermatophytes
microscopic appearance	*Trichophyton mentagrophytes* and *Chrysosporium* spp., which have larger conidia

SEXUAL STATE

None known.

CLINICAL IMPORTANCE

It is a rare cause of nail infection, but is more often encountered as a contaminant of skin and nails.

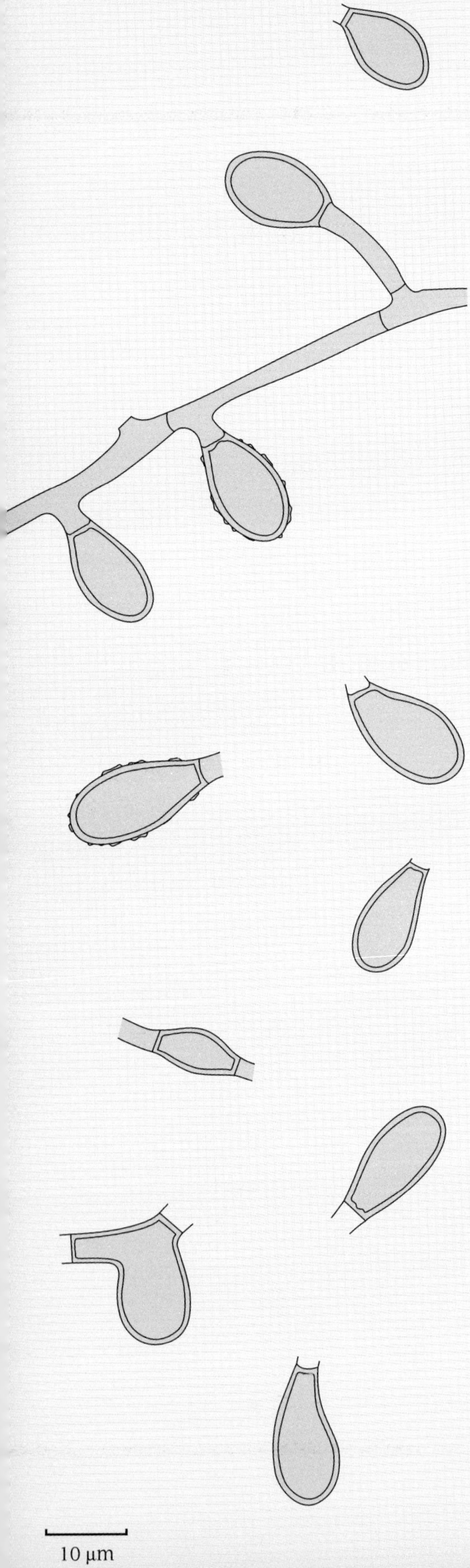

CHRYSOSPORIUM KERATINOPHILUM

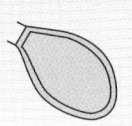

COLONIAL APPEARANCE
at 30°C on glucose peptone agar

diameter	30 mm in one week
topography	flat
texture	powdery to suede
colour	white to cream
reverse	cream

MICROSCOPIC APPEARANCE
at 30°C

predominant features	abundant aleuriospores resembling the microconidia of dermatophyte species but larger
conidia	large, smooth to slightly rough; oval with a truncated base; often septate, intercalary, borne on tips of hyphae or short lateral branches; some spores slightly curved

10 µm

DIFFERENTIAL DIAGNOSIS

colonial appearance   *Trichophyton mentagrophytes*, other white granular dermatophytes

microscopic appearance   *Epidermophyton floccosum, T. tonsurans*

Scedosporium spp., *Myceliophthora thermophila*, other *Chrysosporium* spp. (see *Aphanoascus fulvescens*)

SEXUAL STATE

Aphanoascus keratinophilus

CLINICAL IMPORTANCE

It is a soil organism which has occasionally been isolated from skin and nails, and may be capable of infection.

MYCELIOPHTHORA THERMOPHILA

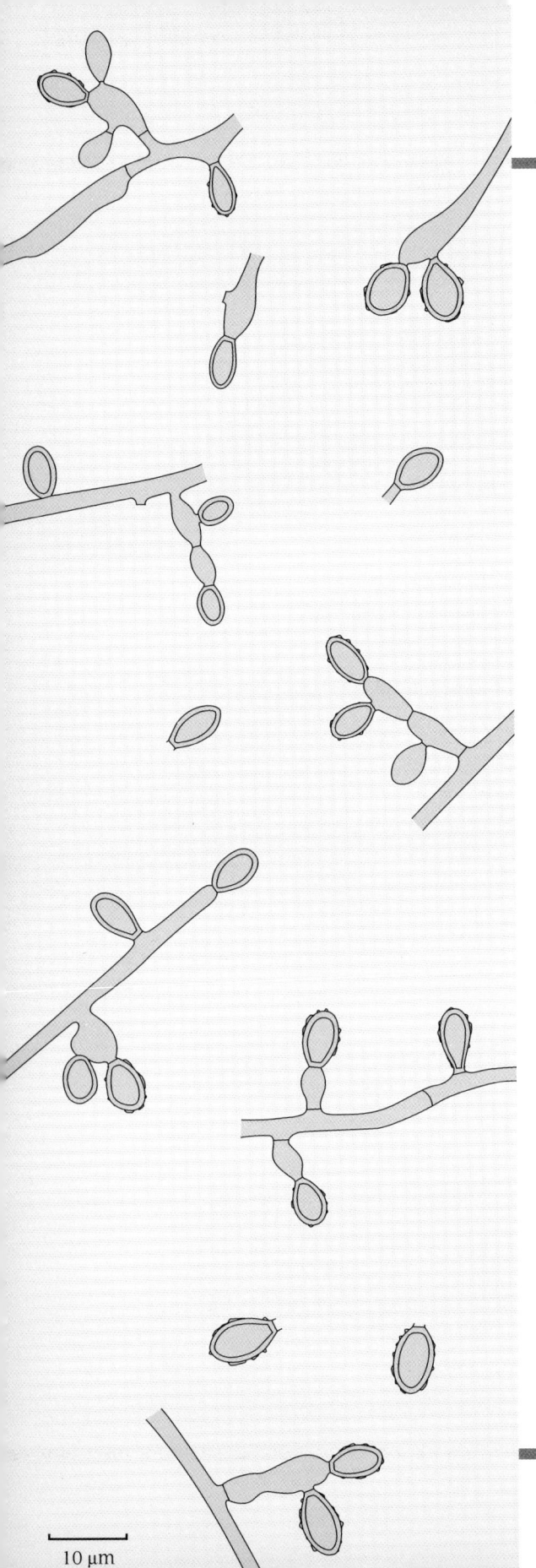

COLONIAL APPEARANCE
at 30°C and 37°C on glucose peptone agar

diameter	60-70 mm in one week
topography	flat
texture	powdery with cotton-like tufts
colour	white with cinnamon-brown centre
reverse	cream

MICROSCOPIC APPEARANCE
at 30°C and 37°C

predominant features	abundant, large aleuriospores borne terminally or laterally on hyphae
conidia	oval to club-shaped, 4.5-11 µm × 3-4.5 µm; hyaline to light brown, smooth or finely roughened, thick-walled; borne on sides or ends of short side-branches, often on a swollen hyphal cell; several conidia may be formed from one cell; occasionally a secondary conidium may be produced from the distal end of the first

DIFFERENTIAL DIAGNOSIS

colonial appearance *Aspergillus terreus, Scopulariopsis brevicaulis, Histoplasma capsulatum*

microscopic appearance *Chrysosporium* spp., but conidia are borne on swollen hyphal cells

SEXUAL STATE

None known.

CLINICAL IMPORTANCE

It is a rare cause of deep infection in immunocompromised patients.

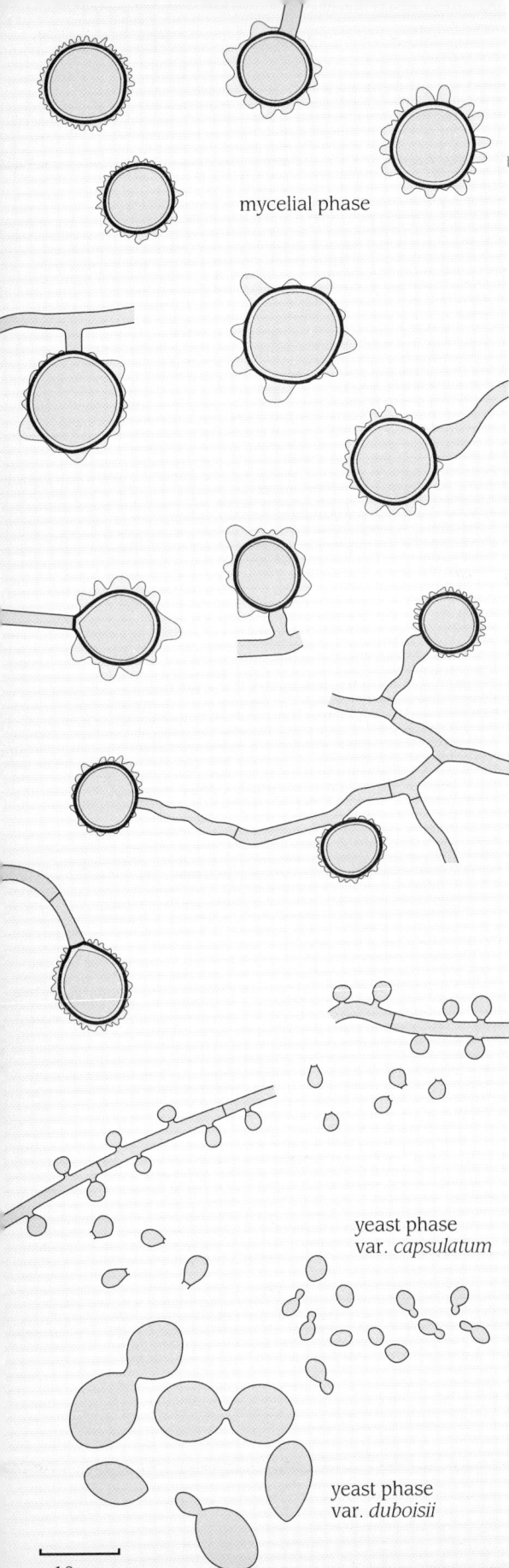

mycelial phase

yeast phase var. *capsulatum*

yeast phase var. *duboisii*

10 μm

HISTOPLASMA CAPSULATUM

(Hazard Group 3 pathogen)

COLONIAL APPEARANCE
at 30°C on glucose peptone agar

diameter	10 mm in one week
topography	flat, heaped in centre
texture	floccose to powdery
colour	white to buff
reverse	buff-brown

at 37°C

topography	yeast-like colonies
texture	rough
colour	cream to brown
reverse	non-pigmented

MICROSCOPIC APPEARANCE
at 30°C

predominant features	large, tuberculate, globose macroconidia
macroconidia	round, 6-15 μm in diameter with spiny or tuberculate walls; borne on short lateral hyphae
microconidia	small, round, sparse or abundant on short, lateral pegs

at 37°C

predominant features	oval yeast cells, 2-3 μm × 3-4 μm, budding on a narrow base (conversion to the yeast form is often incomplete)

VARIANT FORM

var. *duboisii* found in Africa; differs only in having larger yeast cells, 8-15 μm in length with thick walls

DIFFERENTIAL DIAGNOSIS

colonial appearance

- at 30°C   *Blastomyces dermatitidis*, *Chrysosporium* spp., *Renispora* spp. (not described) and *Sepedonium* spp. (not described)

- at 37°C   *B. dermatitidis*, *Trichosporon* spp.

microscopic appearance

- at 30°C   *Myceliophthora* spp., *Renispora* spp., *Sepedonium* spp. (these produce tuberculate macroconidia but not microconidia and do not convert to the yeast form at 37°C)

- at 37°C   *Candida glabrata*; non-encapsulated *Cryptococcus* spp. resemble var. *duboisii*

Note: full identification of *H. capsulatum* requires demonstration of the appropriate exoantigen and/or conversion to the yeast form at 37°C.

SEXUAL STATE

Ajellomyces capsulatus

CLINICAL IMPORTANCE

It is the cause of histoplasmosis in man and mammals. Following inhalation it causes acute pulmonary infection, which may be mild or subclinical but can proceed to cause chronic infection of the lungs or more widespread infection in predisposed individuals. Histoplasmosis has a global distribution, but most cases occur in the central region of the USA and in Central and South America. Other endemic regions include Africa, Australia and parts of East Asia. The organism poses a serious threat to laboratory workers handling live cultures.

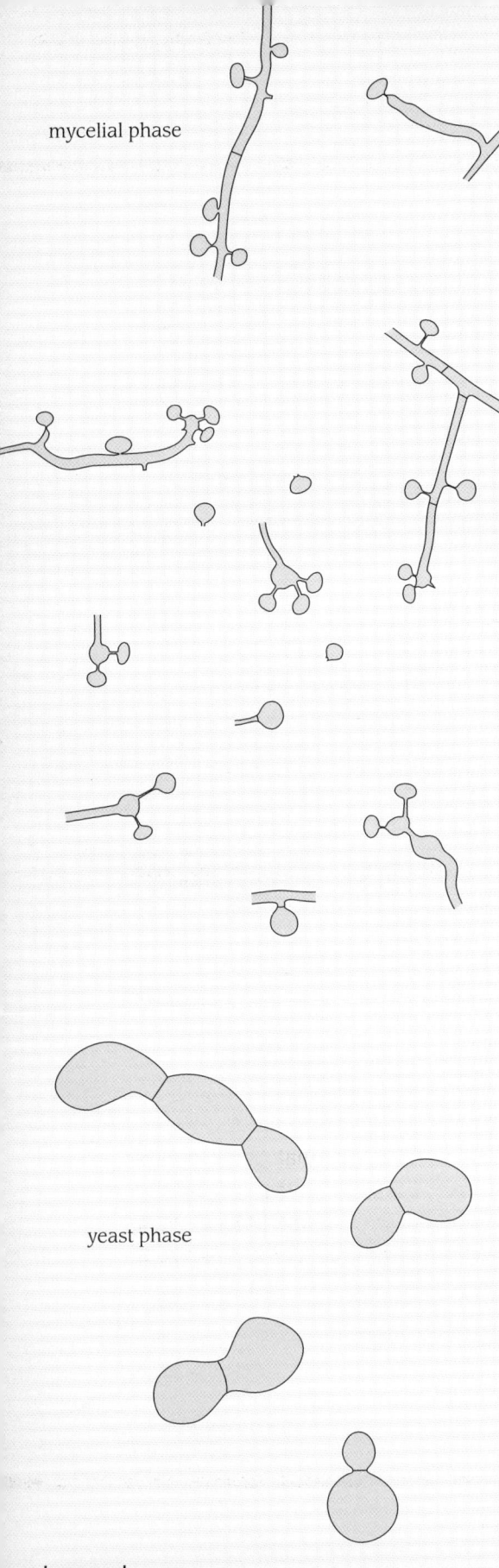

mycelial phase

yeast phase

10 μm

BLASTOMYCES DERMATITIDIS

(Hazard Group 3 pathogen)

COLONIAL APPEARANCE
at 30°C on glucose peptone agar

diameter	10 mm in one week
topography	flat, sometimes folded, with heaped centre
texture	glabrous, with floccose centre
colour	white to buff
reverse	cream to brownish

at 37°C

topography	yeast-like colonies
texture	rough
colour	cream to pale brown
reverse	non-pigmented

MICROSCOPIC APPEARANCE
at 30°C

predominant features	fine hyphae; small round conidia sometimes present
conidia	round to oval, 2-7 μm × 2-4.5 μm; smooth, colourless; borne on very narrow side branches of sometimes swollen hyphal cells or on the ends of longer unbranched hyphae

at 37°C

predominant features	large yeast cells 8-15 μm in diameter with broad-based buds; a septum often separates cells; irregularly shaped pseudohyphal cells may be seen

VARIANT FORM

African form   sporulates more heavily than the American form and at 37°C produces chain-like clusters of yeast cells

DIFFERENTIAL DIAGNOSIS

colonial appearance

- at 30°C   *Paracoccidioides brasiliensis*, *Trichosporon* spp., *Trichophyton schoenleinii*

- at 37°C   *Histoplasma capsulatum*, *P. brasiliensis*, *Trichosporon* spp.

microscopic appearance

- at 30°C   *H. capsulatum* strains that lack macroconidia; *Emmonsia parva* (not described)

- at 37°C   non-budding cells of *P. brasiliensis*, *H. capsulatum* var. *duboisii*, non-encapsulated *Cryptococcus* spp.

Note: full identification of *B. dermatitidis* requires demonstration of the appropriate exoantigen and/or conversion to the yeast form at 37°C.

SEXUAL STATE

Ajellomyces dermatitidis

CLINICAL IMPORTANCE

It is the cause of blastomycosis in man and mammals. Following inhalation it causes pulmonary infection in normal individuals and may spread to other organs, in particular the skin and bones. Most cases occur in the South Central and South East USA, but the disease also occurs in Central and South America and parts of Africa. It poses a serious threat to laboratory workers handling live cultures.

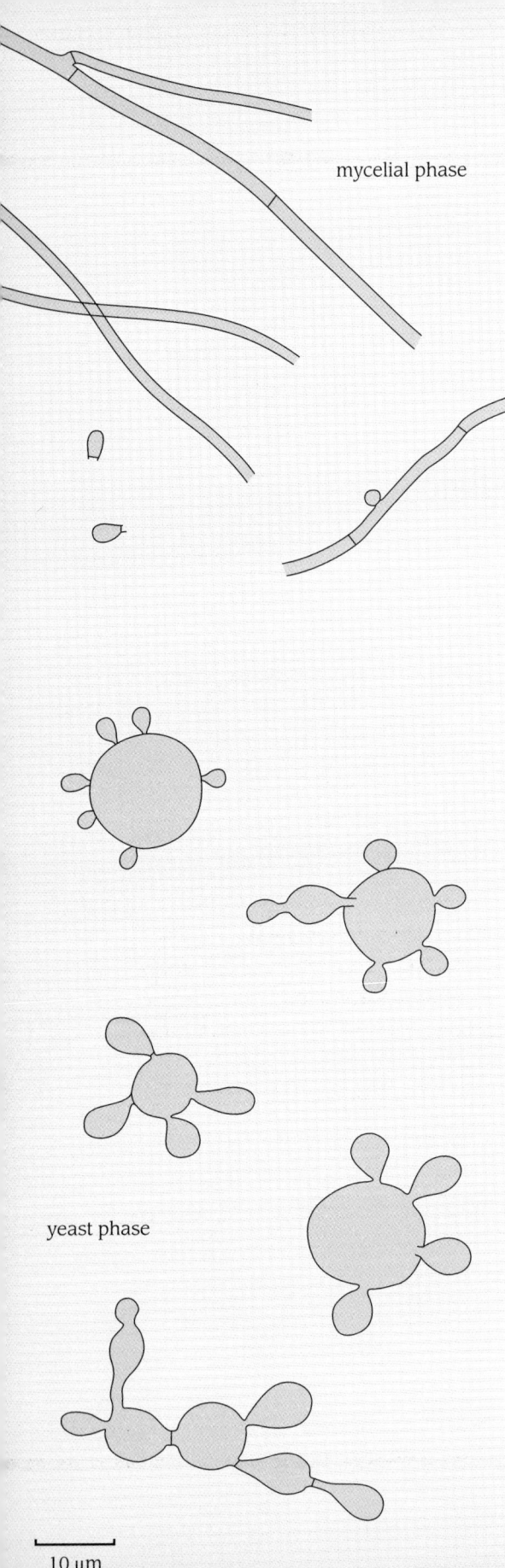

PARACOCCIDIOIDES BRASILIENSIS

(Hazard Group 3 pathogen)

COLONIAL APPEARANCE
at 30°C on glucose peptone agar

diameter	5-10 mm in one week
topography	domed, folded or tufted
texture	floccose, becoming felt-like or velvety
colour	white to buff
reverse	yellow-brown

at 37°C

topography	much folded, heaped, yeast-like colonies
texture	rough
colour	white to cream
reverse	non-pigmented

MICROSCOPIC APPEARANCE
at 30°C

predominant features	hyphae only with occasional chlamydospores
conidia	sporulation poor on glucose peptone agar; on yeast extract agar pear-shaped to oval aleuriospores are produced directly on the sides of the hyphae or on short conidiophores; thick-walled, square to rectangular arthroconidia may also be produced

at 37°C

predominant features	large, round to pear-shaped yeast cells with multiple buds

84

DIFFERENTIAL DIAGNOSIS

colonial appearance

- **at 30°C** *Blastomyces dermatitidis* is similar but grows faster; floccose white dermatophytes; *Trichosporon* spp.

- **at 37°C** *B. dermatitidis*

microscopic appearance

- **at 30°C** any non-sporing white mould

- **at 37°C** *B. dermatitidis, Histoplasma capsulatum* var. *duboisii*, non-encapsulated *Cryptococcus* spp.

Note: full identification of *P. brasiliensis* requires conversion to the yeast form at 37°C.

SEXUAL STATE

None known.

CLINICAL IMPORTANCE

It is the cause of paracoccidioidomycosis in humans. Following inhalation it causes pulmonary infection which is often followed by chronic infection of the buccal, nasal and gastrointestinal mucosa. Most cases have been reported from South and Central America. The organism poses a serious threat to laboratory workers handling live cultures.

6 MOULDS WITH HOLOBLASTIC CO[...]

INTRODUCTION

Almost all the moulds described in this chapter have brown or black pigmented walls and are thus similar in colonial appearance. Conidial morphology and manner of spore production serve as the main distinguishing features. Conidia produced by simple budding (holoblastic conidia) differ widely in morphological type, from minute unicellular spores to massive, thick-walled spores consisting of ten or more cells. In some species the conidia are produced close together on the parent hypha without further development, whereas in others the conidia themselves may bud to produce a chain of spores with the youngest at the tip. In both types further spores are not produced at any point where a conidium has already formed. This is unlike the process involved in enteroblastic conidiation (see Chapters 7 and 8).

For species that form mucoid or wet-surfaced colonies, it is often useful to dissect out material from the edge of the colonies and examine the young, submerged growth. For species forming conidial chains, needle mounts are less satisfactory than adhesive tape preparations or slide culture mounts.

Key to holoblastic moulds with single-celled conidia

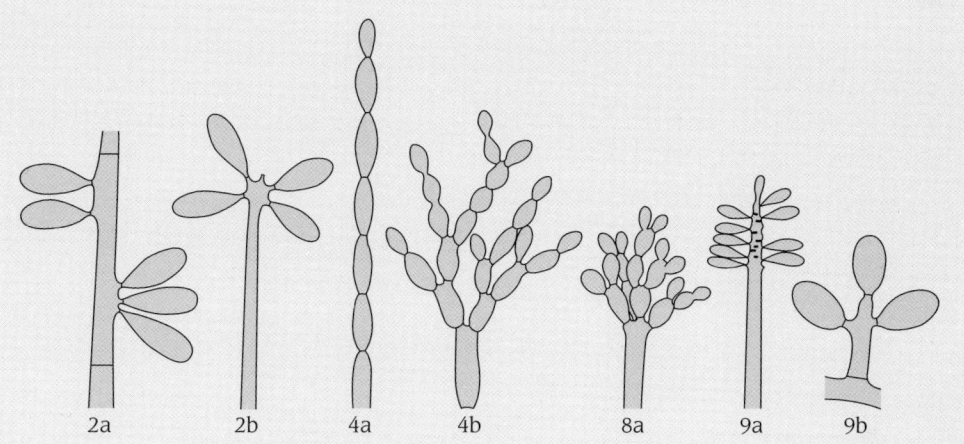

1a	Colony mucoid or glabrous, white or pale pink at first, developing dark colour later	2
1b	Colony floccose or velvety, brown or olive	3
2a	Conidia produced in clusters on the sides of hyphae	*Aureobasidium pullulans*
2b	Conidia produced in terminal rosettes on denticles	*Sporothrix schenckii*
3a	Conidia produced in chains	4
3b	Conidia not in chains	9
4a	Conidial chains long, rarely branched	5
4b	Conidial chains short, much branched	6
5a	No growth or very poor growth at 37°C	*Cladophialophora carrionii*
5b	Good growth at 37°C and 40°C	*Cladophialophora bantiana*
6a	Conidia rough-walled	*Cladosporium herbarum*
6b	Conidia smooth	7
7a	Most conidia round	*Cladosporium sphaerospermum*
7b	Most conidia oval or lemon-shaped	8
8a	Conidial chains very short, compact	*Fonsecaea pedrosoi*
8b	Conidial chains longer, more divergent	*Cladosporium cladosporioides*
9a	Conidia elongate, less than 2 μm wide	*Rhinocladiella atrovirens*
9b	Conidia broadly oval, more than 2 μm wide	*Ramichloridium mackenziei*

Key to holoblastic moulds with multicelled conidia

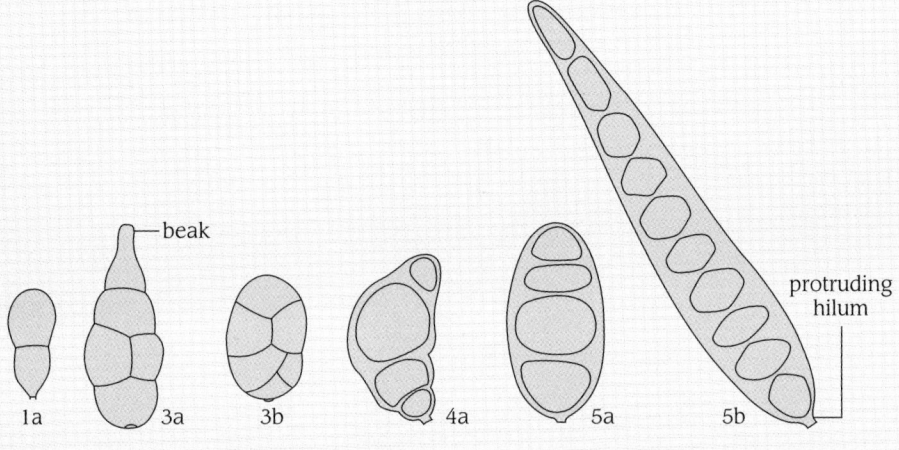

1a	Conidia all two-celled, thin-walled	*Ochroconis gallopava*
1b	Conidia with more than two cells	2
2a	Some conidia with oblique or longitudinal septa in addition to transverse septa	3
2b	Conidia with transverse septa only	4
3a	Conidia in chains, often of more than three; most conidia with apical beak	*Alternaria alternata*
3b	Conidia formed singly or in chains of two or three; most conidia without a beak	*Ulocladium chartarum*
4a	Conidia asymmetrical	*Curvularia lunata*
4b	Conidia oval or cylindrical	5
5a	Conidia oval with few septa, hila not protruding	6
5b	Conidia cylindrical with many septa and strongly protruding hila	7
6a	Conidia mostly with five septa	*Bipolaris hawaiiensis*
6b	Conidia mostly with three septa	*Bipolaris australiensis*
7a	Conidia widest near the base	8
7b	Conidia not widest near the base; some warty projections on conidia walls	*Exserohilum mcginnisii*
8a	Conidia with both the basal and apical septum dark	*Exserohilum rostratum*
8b	Conidia with basal septum only darker than the others	*Exserohilum longirostratum*

AUREOBASIDIUM PULLULANS

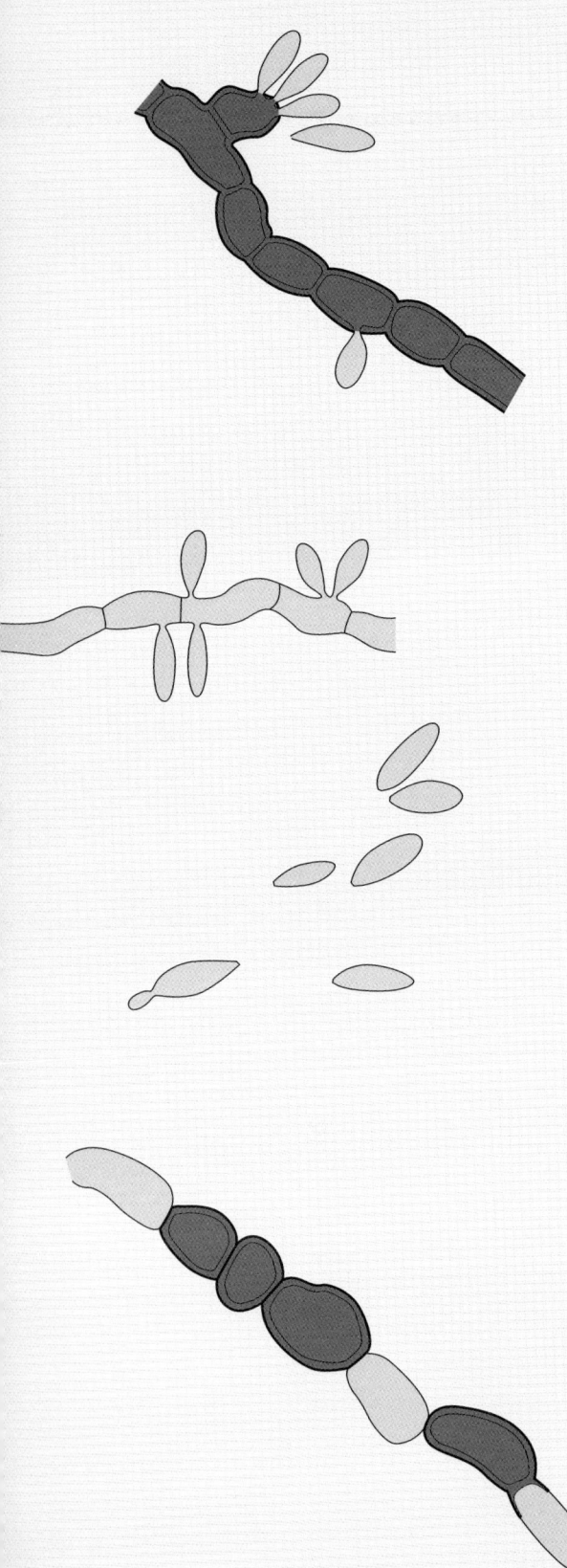

COLONIAL APPEARANCE
at 30°C on glucose peptone agar

diameter	30 mm in one week
topography	flat, spreading
texture	mucoid or glabrous
colour	white to light pink, turning black or dark brown with age
reverse	cream, turning brown or black with age

MICROSCOPIC APPEARANCE
at 30°C

predominant features	yeast-like conidia arising in groups from submerged hyphae; older hyphae develop swollen, black-pigmented cells
conidia	smooth, oval, non-pigmented; produced in clusters from the sides of the hyphae

10 µm

DIFFERENTIAL DIAGNOSIS

colonial appearance   other black yeasts, including *Exophiala* spp., *Lecythophora* spp., *Sporothrix schenckii*

microscopic appearance   *Exophiala* spp., but *A. pullulans* does not produce a succession of spores from the same point

SEXUAL STATE

None known.

CLINICAL IMPORTANCE

This common saprophyte is a rare cause of infection in immunocompromised patients.

SPOROTHRIX SCHENCKII

COLONIAL APPEARANCE
at 30°C on glucose peptone agar

diameter	15 mm in one week
topography	flat to wrinkled
texture	moist, membranous, becoming felt-like
colour	white with grey areas, becoming black
reverse	grey to black
at 37°C	smooth, cream-coloured colonies if conversion to yeast form complete; otherwise similar to that at 30°C but growth is slower

MICROSCOPIC APPEARANCE
at 30°C

predominant features	hyaline hyphae, delicate conidiophores bearing an apical rosette of minute conidia
conidia	oval, 3-6 µm × 2-3 µm, with the proximal end tapered to a point; smooth, thin-walled; formed as a rosette on apex of conidiogenous cell; later-formed conidia are produced as lateral buds and are pigmented
at 37°C	long to oval budding cells, 3-10 µm × 1-3 µm; the yeast form is best produced on special media

yeast phase

VARIANT FORM

var. *lurei* narrower, longer conidia, 3.5-10 µm × 1.5-2 µm; produces thick-walled round cells *in vivo* in addition to the tissue-phase yeast form

DIFFERENTIAL DIAGNOSIS

colonial appearance other *Sporothrix* spp. (these do not convert to the yeast form at 37°C), *Lecythophora mutabilis*, *Aureobasidium pullulans*, *Trichosporon* spp.

microscopic appearance other *Sporothrix* spp., *A. pullulans*

SEXUAL STATE

None known.

CLINICAL IMPORTANCE

It is the cause of sporotrichosis and can cause cutaneous or subcutaneous infection following implantation; lymphatic spread is common. The limbs, and especially the hands and fingers, are the usual sites of infection. The infection can become widespread, particularly in immunocompromised patients. Sporotrichosis has a global distribution.

CLADOPHIALOPHORA CARRIONII

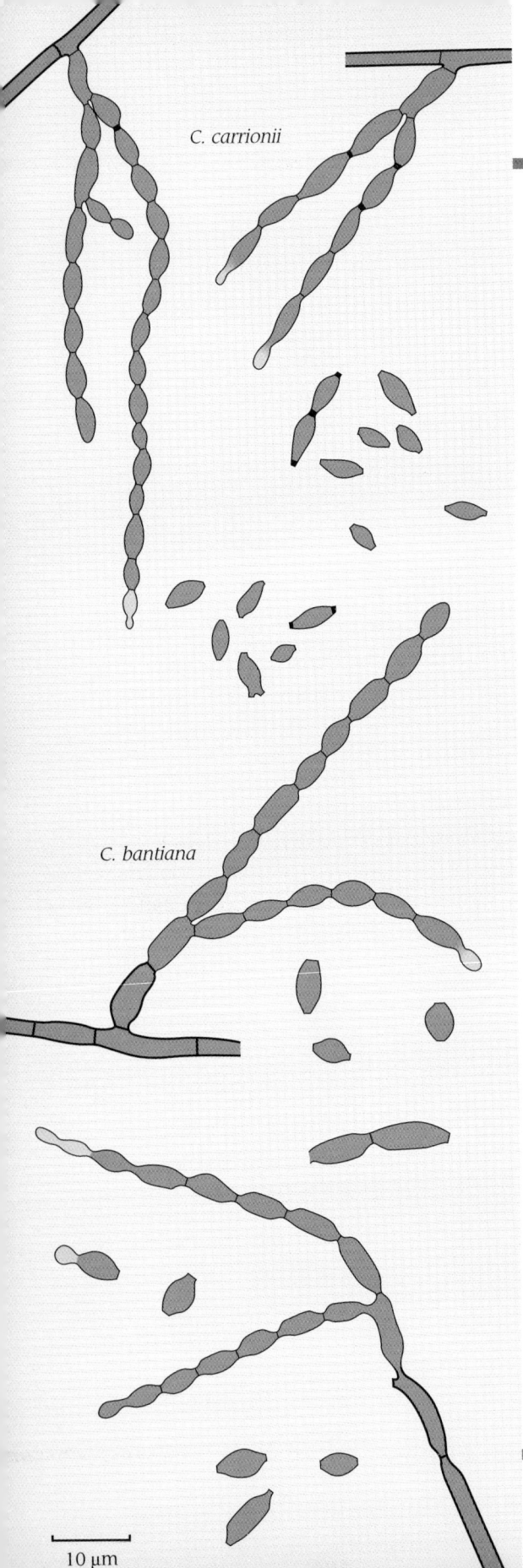

C. carrionii

C. bantiana

10 µm

COLONIAL APPEARANCE
at 30°C on glucose peptone agar

diameter	30 mm in one week
topography	flat or heaped; often irregularly folded in centre
texture	powdery to felt-like
colour	dark olive-green
reverse	olive-black

MICROSCOPIC APPEARANCE
at 30°C

predominant features	long chains of pale brown, lemon-shaped conidia
conidia	pale brown, formed in long, rarely branched chains that are robust and remain intact when mounted; one-celled, smooth or finely roughened, 4.5–8.5 µm × 2.2–2.6 µm, with up to three dark, narrow, inconspicuous scars

DIFFERENTIAL DIAGNOSIS

colonial appearance	other *Cladophialophora* spp., *Cladosporium* spp., *Phialophora* spp., *Fonsecaea* spp., *Exophiala* spp.
microscopic appearance	other *Cladophialophora* spp., especially *C. bantiana*; unlike *C. bantiana*, *C. carrionii* does not grow at 40°C

SEXUAL STATE

None known.

CLINICAL IMPORTANCE

It is a cause of chromoblastomycosis in tropical and subtropical regions.

CLADOPHIALOPHORA BANTIANA

This species grows well at 37-40°C. It has pale, elongated conidia, 6-12.5 μm × 2.5-4 μm, with truncated or pointed ends. Long branching chains of spores that usually remain intact on mounting are present. It is an environmental organism which can cause cerebral phaeohyphomycosis.

CLADOSPORIUM SPHAEROSPERMUM

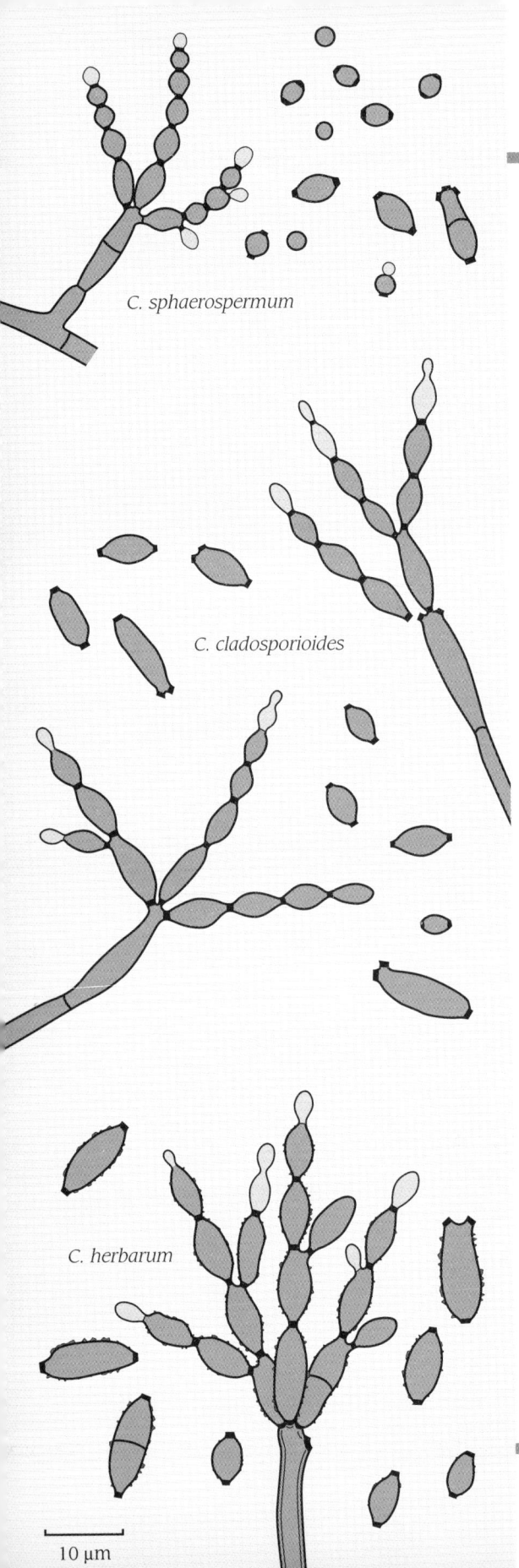

C. sphaerospermum

C. cladosporioides

C. herbarum

10 µm

COLONIAL APPEARANCE
at 30°C on glucose peptone agar

diameter	10 mm in one week
topography	domed, often folded
texture	powdery
colour	dark grey-green
reverse	green-black with a paler edge

MICROSCOPIC APPEARANCE
at 30°C

predominant features	branching chains of dark, round spores forming tree-like structures
conidia	dark brown; round or nearly so, 3-4.5 µm in diameter; mostly single-celled, with two prominent scars; chains formed by budding with the youngest spore at the tip; older conidia at the base are oval or shield-shaped, up to 15 µm in length, often septate and may have several conspicuous scars; chains often break up on mounting

DIFFERENTIAL DIAGNOSIS

colonial appearance   other *Cladosporium* spp., *Phialophora* spp., *Fonsecaea* spp., *Exophiala* spp.

microscopic appearance   other *Cladosporium* spp. (see below), *Cladophialophora* spp.

SEXUAL STATE

None known.

CLINICAL IMPORTANCE

This common environmental organism is a doubtful cause of nail and skin infection.

CLADOSPORIUM HERBARUM

The pronounced roughening on the walls of the conidia of this environmental organism distinguishes it from other *Cladosporium* spp.

CLADOSPORIUM CLADOSPORIOIDES

This common environmental organism can be distinguished from *C. herbarum* by its smooth conidia, and from *C. sphaerospermum* by its larger, lemon-shaped conidia.

FONSECAEA PEDROSOI

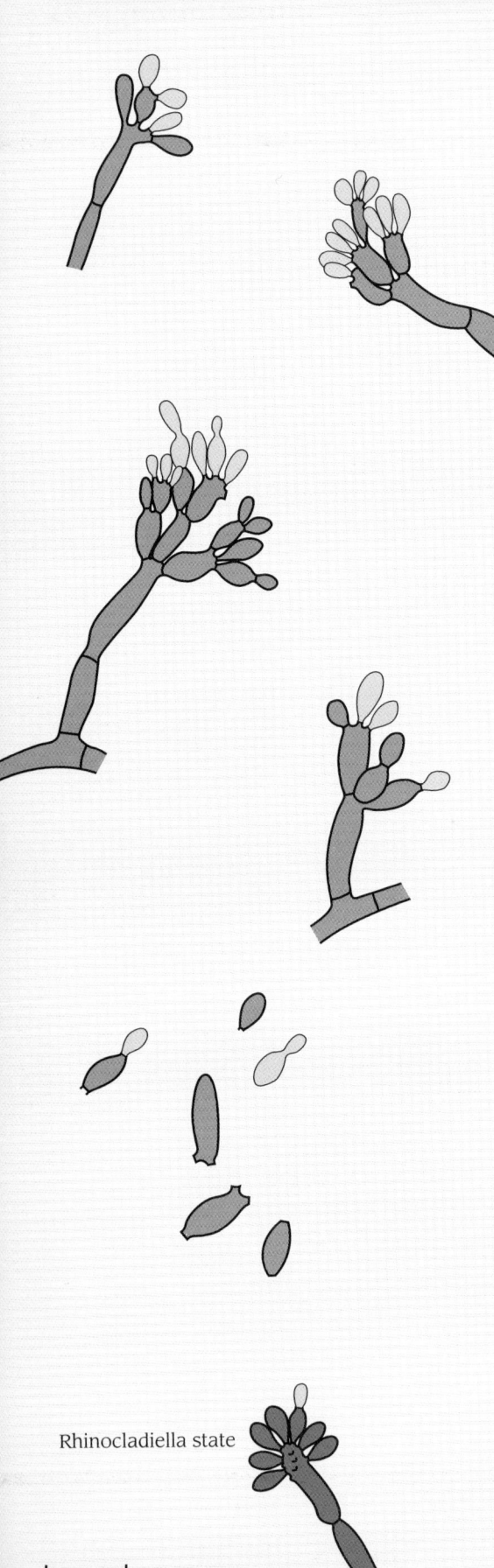

Rhinocladiella state

10 μm

COLONIAL APPEARANCE
at 30°C on glucose peptone agar

diameter	10 mm in one week
topography	heaped in centre with flat edge
texture	floccose with submerged margin
colour	dark brown to olive-black
reverse	black

MICROSCOPIC APPEARANCE
at 30°C

predominant features	brown, one-celled conidia in short, branched chains
conidia	short chains, elliptical to oval, 1.5-3 μm × 3-6 μm, produced by successive budding; in addition, brown, one-celled conidia may be produced on short denticles at the tip of a conidiogenous cell (Rhinocladiella state); most isolates can be induced to produce flask-shaped phialides of the Phialophora type

DIFFERENTIAL DIAGNOSIS

colonial appearance   *Cladosporium* spp., *Phialophora* spp., *Exophiala* spp.

microscopic appearance   *Cladosporium* spp., *Phialophora* spp., *Rhinocladiella* spp.

SEXUAL STATE

None known.

CLINICAL IMPORTANCE

It is the most common cause of chromoblastomycosis. Disseminated infection has been reported.

RHINOCLADIELLA ATROVIRENS

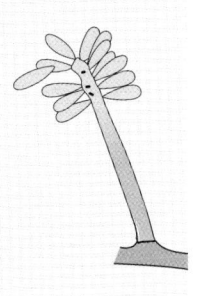

COLONIAL APPEARANCE
at 30°C on glucose peptone agar

diameter	10 mm in one week
topography	flat or heaped
texture	floccose
colour	olive-green to dark grey
reverse	grey to black

MICROSCOPIC APPEARANCE
at 30°C

predominant features	brown conidiophores bearing small, pale brown or colourless conidia towards tip
conidia	small, 4-5.5 µm × 1-2 µm, long to oval; pale brown with inconspicuous scars; formed on apical region of conidiophores which are paler in colour at the tips

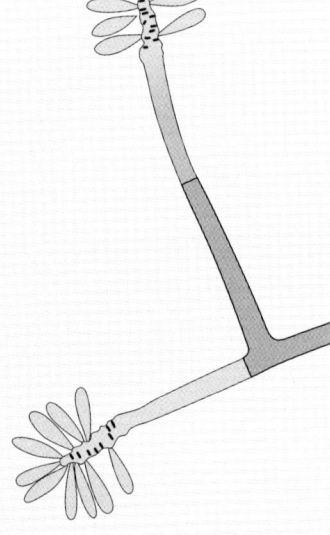

10 µm

DIFFERENTIAL DIAGNOSIS

colonial appearance   *Cladosporium* spp., *Phialophora* spp., *Exophiala* spp., *Fonsecaea* spp.

microscopic appearance   other *Rhinocladiella* spp.; *Fonsecaea pedrosoi* strains with Rhinocladiella-type conidiophores

SEXUAL STATE

None known.

CLINICAL IMPORTANCE

It occurs in the environment and is a rare cause of phaeohyphomycosis in humans.

RAMICHLORIDIUM MACKENZIEI

COLONIAL APPEARANCE
at 30°C on glucose peptone agar

diameter	5 mm in one week
topography	high-domed centre with submerged margin
texture	densely floccose
colour	dark grey-brown to black
reverse	black

MICROSCOPIC APPEARANCE
at 30°C

predominant features	smooth, pigmented, septate hyphae with brown, oval conidia
conidia	brown, oval, 4.7-9.6 µm × 2.7-6.0 µm, with flat secession scar or protuberant hilum; produced from poorly differentiated conidiogenous cells; few conidia produced on each fertile axis; some strains show poor or no sporulation on first isolation but develop abundant conidia on successive subcultures

DIFFERENTIAL DIAGNOSIS

colonial appearance   other *Ramichloridium* spp., *Cladosporium* spp., *Exophiala* spp., *Phialophora* spp., *Rhinocladiella* spp., *Fonsecaea pedrosoi*

microscopic appearance   *F. pedrosoi*, *Rhinocladiella* spp.

SEXUAL STATE

None known.

CLINICAL IMPORTANCE

A number of cases of phaeohyphomycosis of the brain have been described from the Middle East.

OCHROCONIS GALLOPAVA

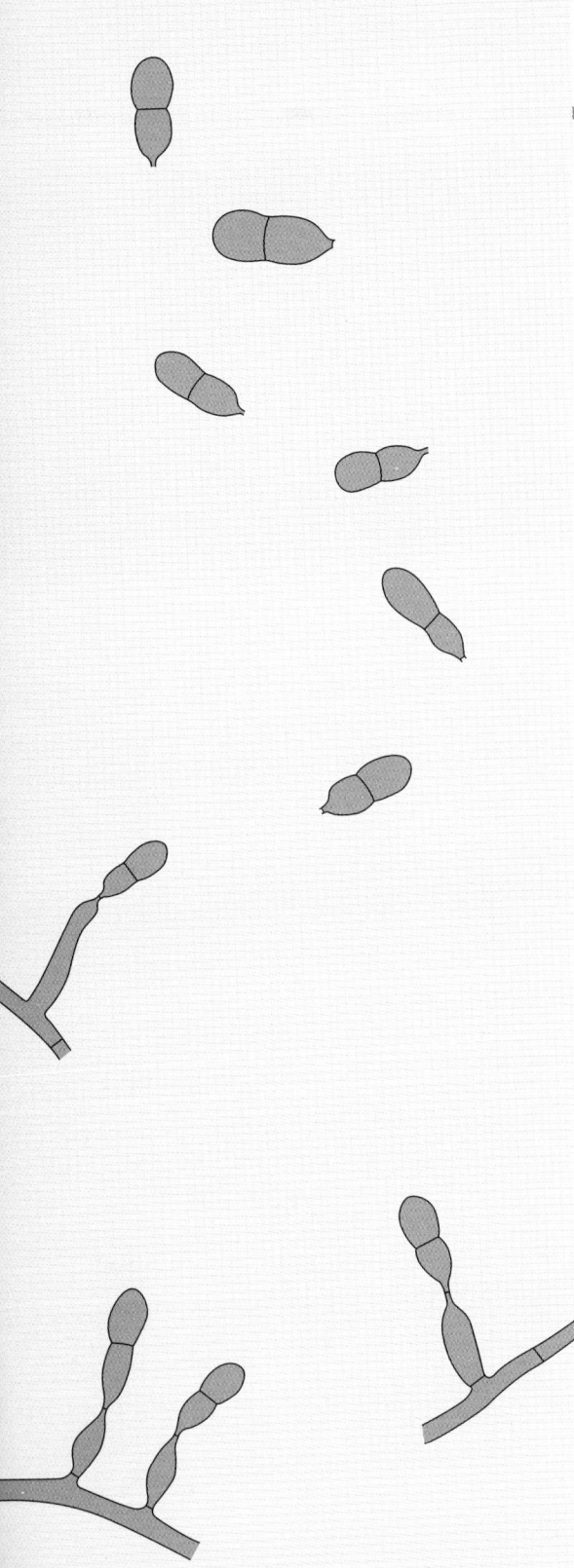

COLONIAL APPEARANCE
at 30°C on glucose peptone agar

diameter	15 mm in one week
topography	flat
texture	granular to velvet
colour	dark red-brown
reverse	reddish brown with diffusible pigment

MICROSCOPIC APPEARANCE
at 30°C

predominant features	pale brown hyphae with small, pale brown, two-celled conidia; thermotolerant
conidia	produced on narrow cylindrical denticles on short side branches of normal hyphae; pale brown, oval to cylindrical, 6-17 µm × 2.5-4.5 µm; two-celled with the apical cell larger, often constricted at the central septum; smooth or roughened with a prominent basal scar

DIFFERENTIAL DIAGNOSIS

Few other fungi have the dark russet colour and granular surface of this species, nor the distinctive, two-celled conidia.

SEXUAL STATE

None known.

CLINICAL IMPORTANCE

This species, which occurs in the environment, is a rare cause of subcutaneous and deep infection in humans.

ALTERNARIA ALTERNATA

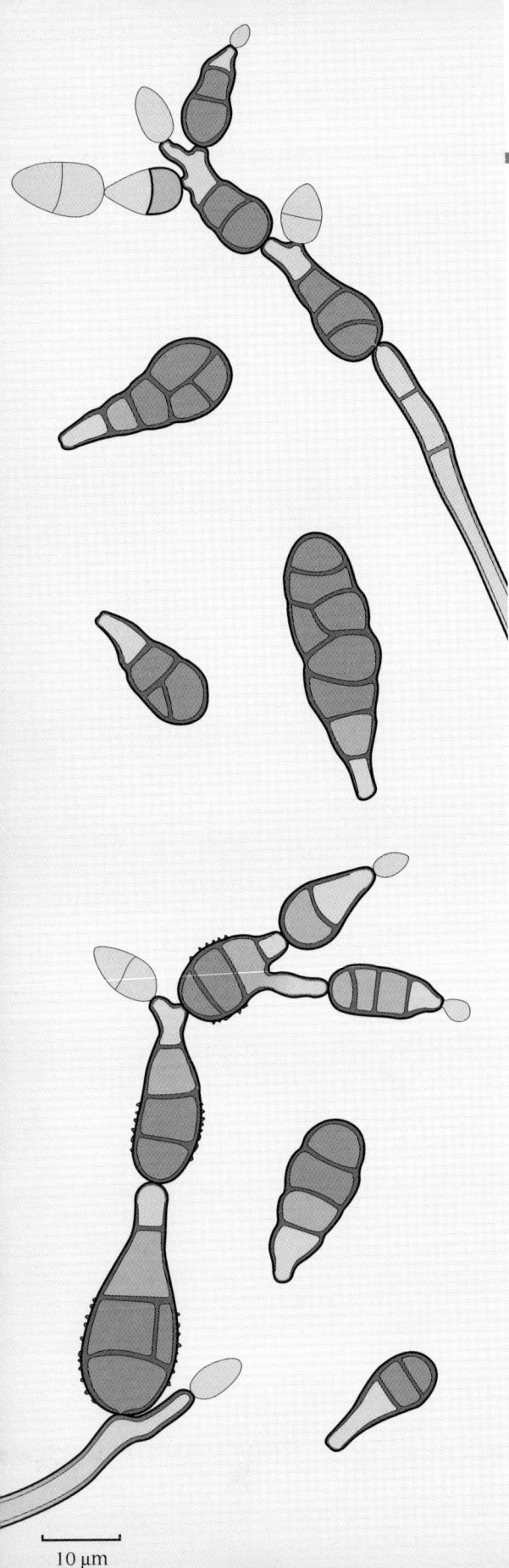

COLONIAL APPEARANCE
at 30°C on glucose peptone agar

diameter	60 mm in one week
topography	flat
texture	powdery to felt-like or floccose
colour	white at first, becoming grey to dark green-black
reverse	cream, becoming grey to black

MICROSCOPIC APPEARANCE
at 30°C

predominant features	chains of pale brown, club-shaped conidia with transverse and longitudinal septa
conidia	club-shaped, pale brown, 20–63 µm × 9–18 µm; produced in long chains; multi-septate with both transverse and longitudinal or oblique septa; mostly with a terminal beak at the distal end

DIFFERENTIAL DIAGNOSIS

colonial appearance *Curvularia* spp., *Ulocladium* spp.

microscopic appearance *Ulocladium* spp., except chains of more than three spores are uncommon in these and most conidia do not possess a beak

SEXUAL STATE

None known.

CLINICAL IMPORTANCE

It is an environmental saprophyte that has caused cutaneous and deep-seated infection in immunocompromised patients.

ULOCLADIUM CHARTARUM

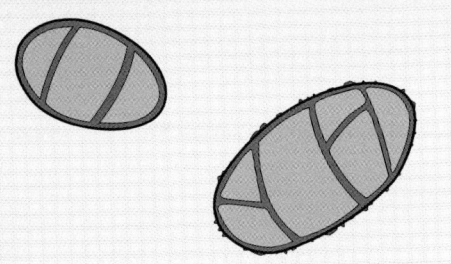

COLONIAL APPEARANCE
at 30°C on glucose peptone agar

diameter	60 mm in one week
topography	flat to spreading
texture	powdery to felt-like
colour	grey-black
reverse	black

MICROSCOPIC APPEARANCE
at 30°C

predominant features	dark brown, oval conidia with transverse and longitudinal septa produced on geniculate conidiophores
conidia	dark brown, oval, multi-septate (one to five septa), with transverse, longitudinal and oblique septa; smooth or rough walls; some conidia may form a short beak and produce a secondary conidium

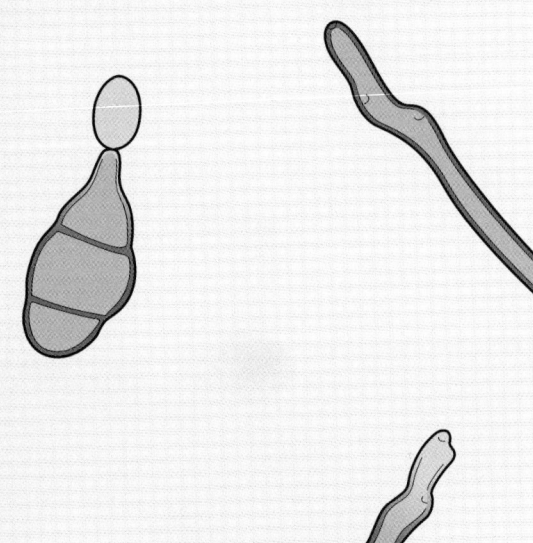

10 µm

DIFFERENTIAL DIAGNOSIS

colonial appearance *Alternaria* spp., *Curvularia* spp., *Bipolaris* spp.

microscopic appearance *Alternaria* spp., except these form conidia in long chains, mostly with prominent beaks

SEXUAL STATE

None known.

CLINICAL IMPORTANCE

It is a common saprophyte, but has been implicated in at least one subcutaneous infection in humans.

CURVULARIA LUNATA

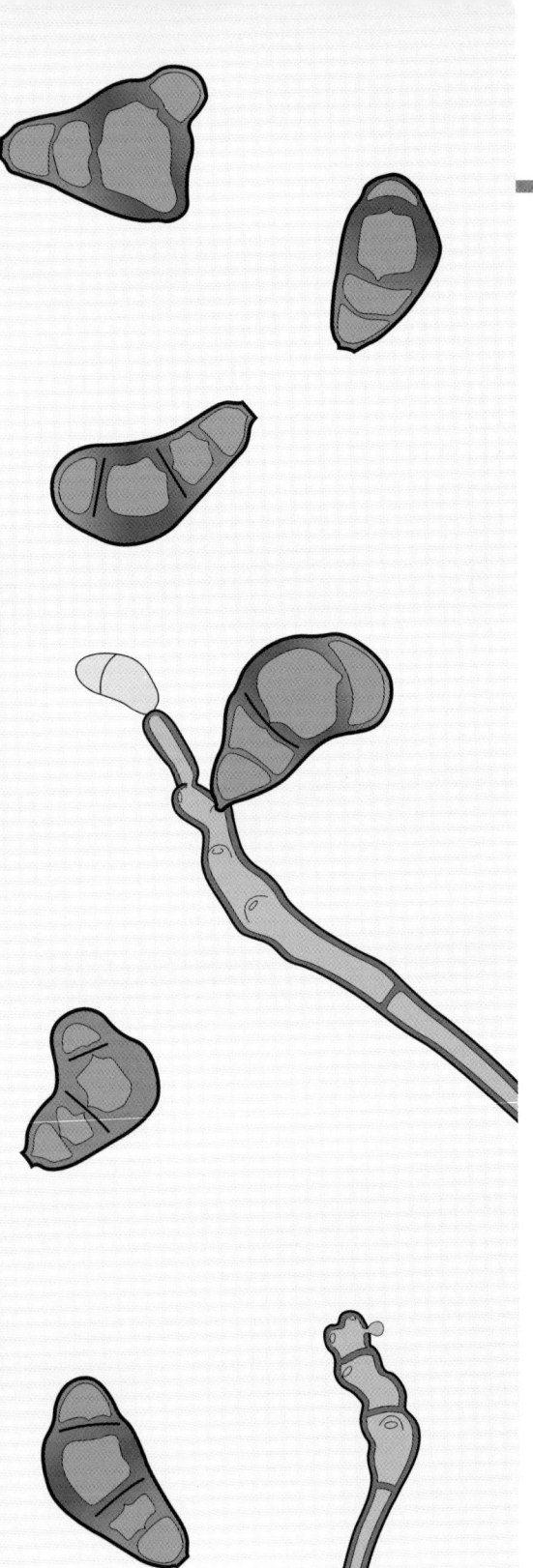

COLONIAL APPEARANCE
at 30°C on glucose peptone agar

diameter	50 mm in one week
topography	flat
texture	floccose to felt-like
colour	dark brown to grey-black, often with a white edge
reverse	dark brown

MICROSCOPIC APPEARANCE
at 30°C

predominant features	dark brown, curved conidia with three septa
conidia	curved, 20-32 µm × 9-15 µm; produced on geniculate conidiophores; typically with three septa; the penultimate cell is asymmetrical and usually darker than the others

DIFFERENTIAL DIAGNOSIS

colonial appearance	*Alternaria* spp., *Ulocladium* spp., *Bipolaris* spp., *Exserohilum* spp.
microscopic appearance	*Bipolaris* spp. and *Exserohilum* spp. (except that conidia are usually curved in *C. lunata*), *Alternaria* spp.

SEXUAL STATE

Cochliobolus lunatus

CLINICAL IMPORTANCE

It is an important plant pathogen in the tropics. In humans it causes keratitis and sinusitis and deep infection in immunocompromised patients has been reported.

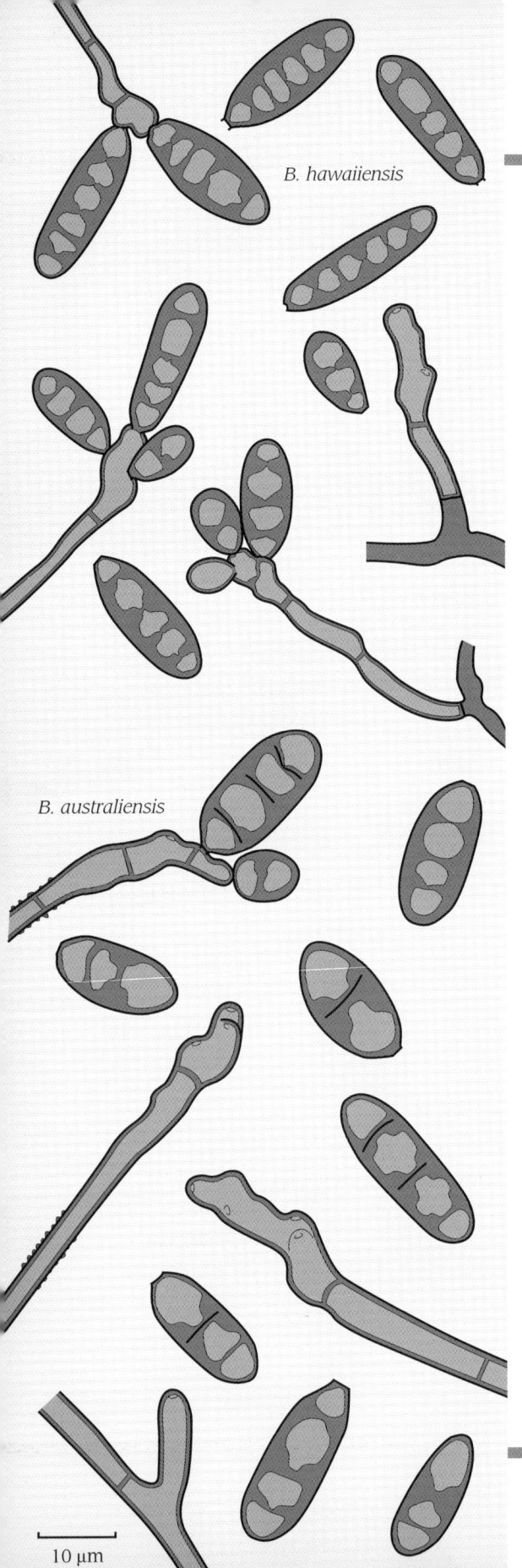

B. hawaiiensis

B. australiensis

10 μm

BIPOLARIS HAWAIIENSIS

COLONIAL APPEARANCE
at 30°C on glucose peptone agar

diameter	80 mm in one week
topography	flat, spreading
texture	floccose to felt-like
colour	grey-black, often with white edge
reverse	black

MICROSCOPIC APPEARANCE
at 30°C

predominant features	large, oval, brown-pigmented, thick-walled, multi-septate conidia produced on geniculate conidiophores
conidia	large, oval to elliptical, rounded at both ends, 16-34 μm × 4-9 μm, with up to six septa (typically four or five)

DIFFERENTIAL DIAGNOSIS

colonial appearance   other *Bipolaris* spp., *Exserohilum* spp., *Alternaria* spp., *Curvularia* spp., *Ulocladium* spp.

microscopic appearance   other *Bipolaris* spp., which differ from *B. hawaiiensis* in conidial form, size and number of septa

Exserohilum spp., which form protruding truncate hila

Helminthosporium spp. (not described), but conidiophores of this genus are not geniculate

Drechslera spp. (not described), except these germinate at right angles to the spore axis

Curvularia spp., which have curved conidia

SEXUAL STATE

Cochliobolus hawaiiensis

CLINICAL IMPORTANCE

It is a tropical plant pathogen and a rare cause of deep infection, usually in immunocompromised patients.

BIPOLARIS AUSTRALIENSIS

This organism closely resembles *B. hawaiiensis* both in colonial and microscopic appearance, except that its conidia generally have fewer septa. Typically conidia from isolates of *B. hawaiiensis* have five septa whereas 80-90% of those of *B. australiensis* have three and occasionally four or five. It is a plant pathogen, but deep infection in humans has been reported.

EXSEROHILUM ROSTRATUM

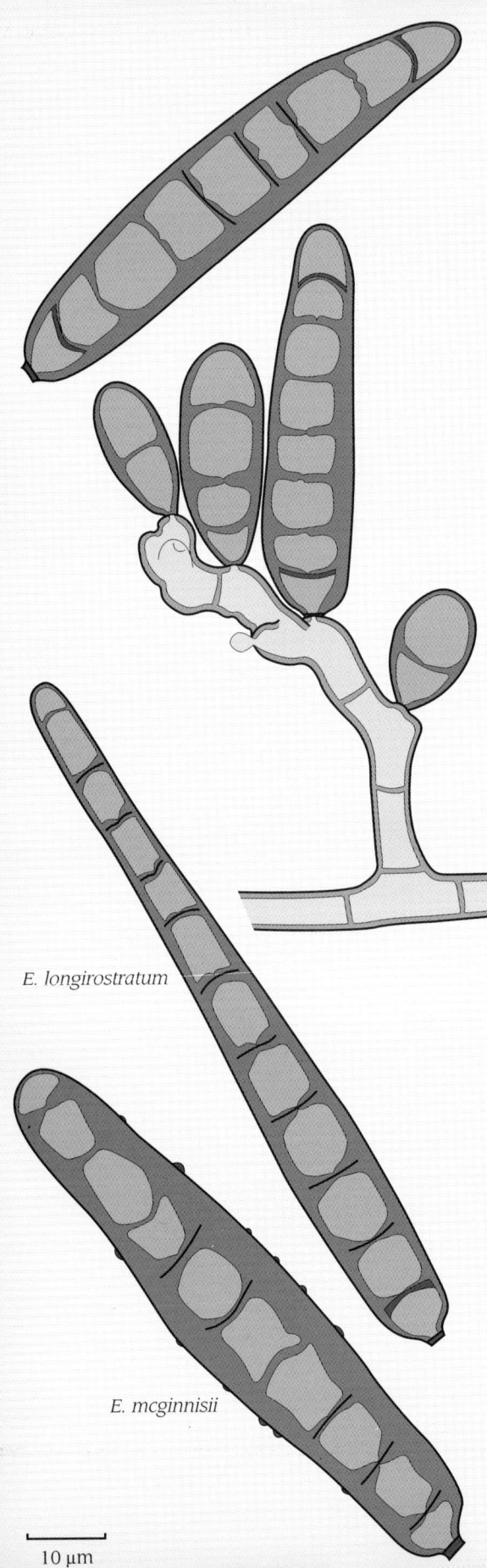

E. longirostratum

E. mcginnisii

10 μm

COLONIAL APPEARANCE
at 30°C on glucose peptone agar

diameter	80 mm in one week
topography	flat
texture	floccose to felt-like
colour	grey-black, often with white edge
reverse	black

MICROSCOPIC APPEARANCE
at 30°C

predominant features	large, brown-pigmented, thick-walled, multi-septate conidia produced on geniculate conidiophores
conidia	large, mostly straight, ellipsoidal to elongate, 30-128 μm × 9-23 μm, mostly with seven to nine septa; the septa immediately above the hilum and near the tip of the conidium are thicker and darker; basal scar (hilum) is prominent and protrudes

DIFFERENTIAL DIAGNOSIS

colonial appearance   other *Exserohilum* spp., *Bipolaris* spp., *Alternaria* spp., *Curvularia* spp., *Ulocladium* spp.

microscopic appearance   *Bipolaris* spp., which have less prominent hila and oval, rather than elongate, conidia; *Helminthosporium* spp., but conidiophores of this genus are not geniculate

SEXUAL STATE

Setosphaeria rostrata

CLINICAL IMPORTANCE

This tropical plant pathogen is a rare cause of subcutaneous and deep infection, usually in immunocompromised patients.

EXSEROHILUM LONGIROSTRATUM

This species closely resembles *E. rostratum* both in colonial and microscopic appearance but differs by producing a mix of long and short conidia. Long conidia (196-260 μm × 13-16 μm) typically have 13-21 septa; short conidia (38-79 μm × 13-19 μm) typically have five to nine septa. Occasional cases of human infection have been reported.

EXSEROHILUM MCGINNISII

This species is similar to other *Exserohilum* spp. in appearance. Typical conidia (64-100 μm × 10-15 μm) have warty projections on the walls, mostly with nine to eleven septa; they do not show the darkened thickening of the septum above the hilum that is seen in other *Exserohilum* spp. It is a rare cause of sinusitis.

7 MOULDS WITH ENTEROBLASTIC C

INTRODUCTION

The moulds described in this chapter are characterised by formation of chains of dry conidia from phialidic cells. Chain formation is a more or less constant feature and is simple to recognise. The species described in Chapter 8 also produce spores from phialidic cells but do not exhibit chain formation. No distinction has been made between true phialides and annellides, which are now known to be slight modifications of the enteroblastic spore production process.

Dry spores are adapted for spread by air currents and are very easily dislodged from a growing culture. Thus, care must be taken when handling sporing fungi to avoid inhalation of potential pathogens and contamination of other cultures. Owing to the heavy sporulation of these moulds, their microscopic structure is best studied by examining young conidiophores from near the edge of the colonies. If conidiophores cannot be clearly seen among the mass of conidia, the cover slip should be removed and the specimen transferred to a fresh drop of mounting fluid on a second slide.

Aspergillus and *Penicillium* species are mainly distinguished by the presence in the former genus of the vesicle, a swollen tip to the conidiophore stalk, upon which the phialides are mounted. A few *Aspergillus* species produce non-vesiculate, Penicillium-like heads in addition to the more normal vesiculate heads. Also, some monoverticillate *Penicillium* species produce a very small vesicle. Unlike those of *Aspergillus* species, however, the width of these vesicles is never more than about twice that of the stalk.

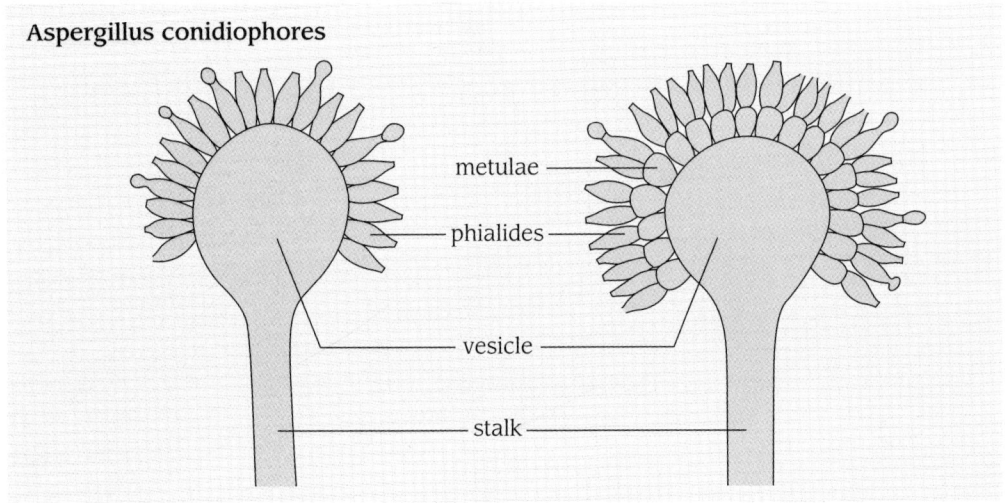

IDIA ADHERING IN CHAINS

The shape of the spore mass before disturbance provides a useful indicator in Aspergillus identification. It is best seen using a plate microscope with top lighting to examine the growing colonies. Several species have a columnar head of spores, but many have a globose, irregular or radiate head. This feature is exhibited only in a mature, well-sporing culture, in many species a week or more old. Another useful feature is the presence of roughened conidiophore stalks. The roughness can be quite fine and high optical contrast microscopy may be needed to observe it. The character that gives most trouble in Aspergillus identification is the presence or absence of metulae, a layer of cells between the vesicle and the spore-forming phialides. The presence of metulae may be detected by squashing the sporing heads almost to destruction under the cover slip and examining the fragmented remains.

The *Aspergillus* species described in this chapter are those more commonly encountered in clinical specimens. However, others, for which no descriptions are given here, have been recorded as rare causes of infection, and some have been encountered as contaminants; the names of several of these organisms are given in the keys that follow. If an isolate cannot be identified by the descriptions given in this chapter, reference should be made to specialist texts. With developments in patient care, a wider range of species is now being reported as significant causes of deep infection.

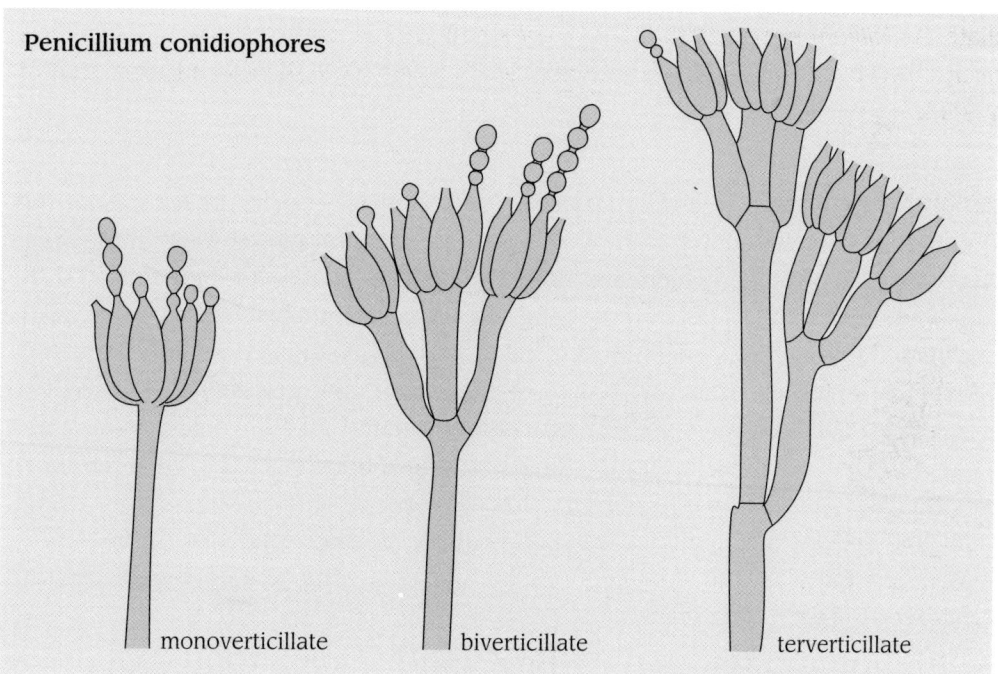

Penicillium conidiophores

monoverticillate   biverticillate   terverticillate

Key to *Aspergillus* species

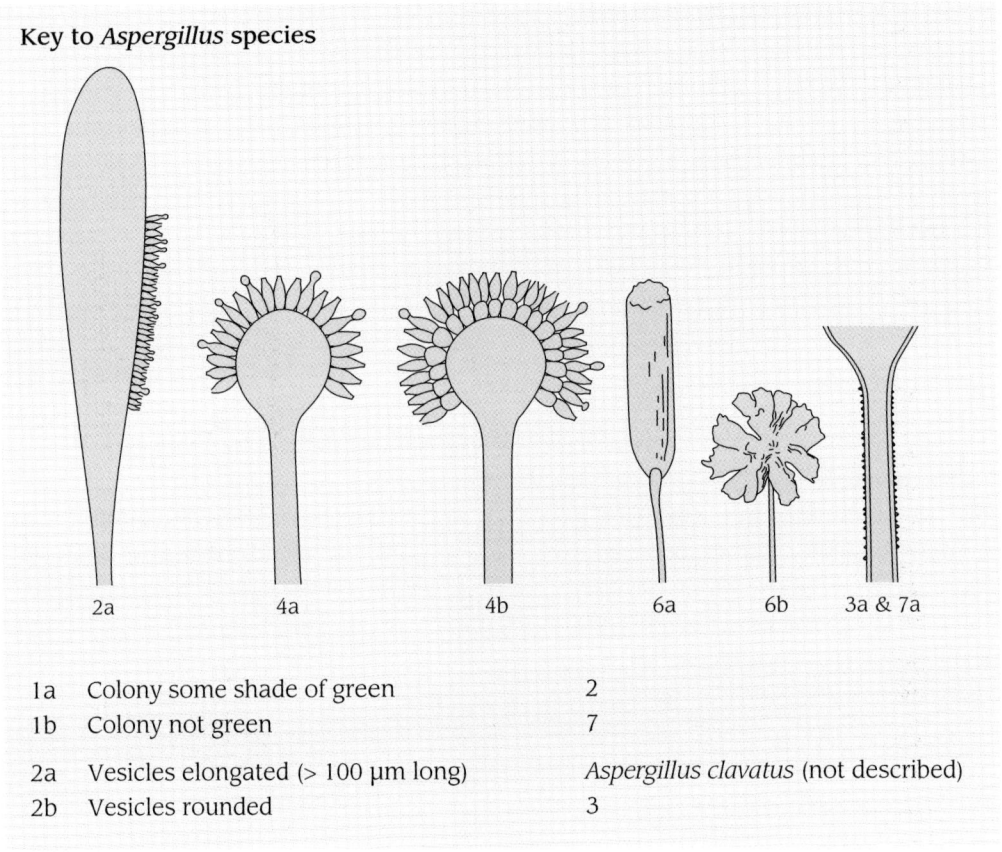

1a Colony some shade of green	2
1b Colony not green	7
2a Vesicles elongated (> 100 µm long)	*Aspergillus clavatus* (not described)
2b Vesicles rounded	3

Although this manual does not describe the numerous species of *Penicillium* that laboratories may encounter as contaminants, the following description of the major morphological groups, which are based upon the degree of branching of the conidiophore within the genus, may be useful. In monoverticillate species the stalk is crowned by a number of phialides, i.e. there is a single point of branching. Biverticillate species have two points of branching, and are subdivided into two groups: those with very symmetrical branching systems and elongated phialides, and those with less symmetrical branching and short phialides. The large group of terverticillate *Penicillium* species are characterised by three points of branching. Individual species are identified by such features as shape and size of the various component parts of the conidiophore and conidia. The reader should refer to specialist texts for full speciation.

The Hazard Group 3 pathogen *Penicillium marneffei* is a biverticillate species which

3a	Colony uniformly yellow-green, stalks rough	*Aspergillus flavus*
3b	Colony dark green, sometimes with yellow areas; stalks smooth	4
4a	Metulae absent	5
4b	Metulae present	6
5a	Phialides on upper two thirds of small flattened vesicle	*Aspergillus fumigatus*
5b	Phialides over entire surface of large round vesicle	*Aspergillus glaucus*
6a	Stalks pale brown, heads columnar in old cultures	*Aspergillus nidulans*
6b	Stalks colourless, heads globose or irregular	*Aspergillus versicolor*
7a	Stalks rough, colony orange-brown	*Aspergillus ochraceus* (not described)
7b	Stalks coloured brown or yellow	8
7c	Stalks colourless	9
8a	Colony dull grey to charcoal	*Aspergillus ustus*
8b	Colony yellow to buff	*Aspergillus flavipes* (not described)
9a	Colony black or dark brown	*Aspergillus niger*
9b	Colony cinnamon-brown to sand coloured	*Aspergillus terreus*
9c	Colony white or pale cream	*Aspergillus candidus*

Key to other species

1a	Colony green	*Penicillium* spp. (but see 3b)
1b	Colony not green	2
2a	Conidia large, round, with a flat scar	*Scopulariopsis brevicaulis*
2b	Conidia small, oval; scar minute	3
3a	Colony pale purple	*Paecilomyces lilacinus*
3b	Colony greenish buff	*Paecilomyces variotii*

produces a diffusing, red pigment. It is, however, not the only species of *Penicillium* with this combination of features. Suspect isolates should always be referred to a specialist laboratory.

ASPERGILLUS FLAVUS

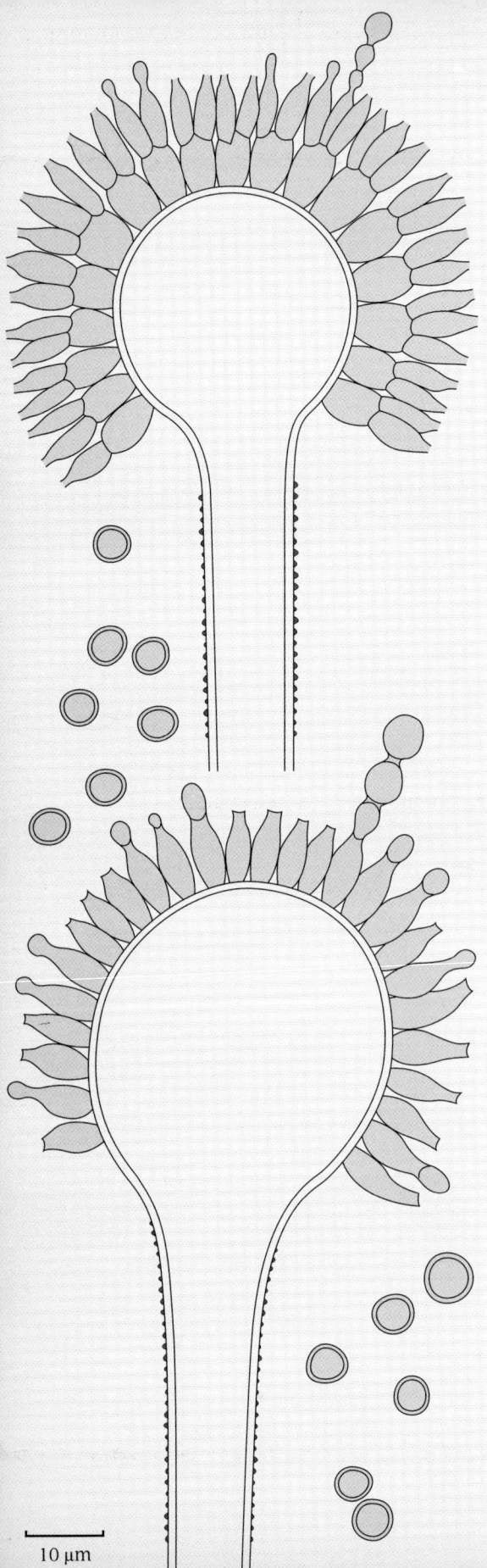

COLONIAL APPEARANCE
at 30°C on glucose peptone agar

diameter	60 mm in one week
topography	flat
texture	floccose to granular
colour	bright yellow-green; occasionally yellow-brown
reverse	cream

MICROSCOPIC APPEARANCE
at 30°C

predominant features	vesiculate conidiophores; numerous, relatively large conidia
conidiophore	roughened stalks; vesicles globose with radiate or columnar spore production; phialides arising directly from the entire surface of the vesicle in some heads, and produced on metulae in others
conidia	round to elliptical, 3-6 µm; smooth or finely roughened

DIFFERENTIAL DIAGNOSIS

colonial appearance *Aspergillus parasiticus* (not described); *A. oryzae* (not described) has a stronger olive colour

microscopic appearance *A. parasiticus* has no metulae and rough conidia

A. oryzae has heads with metulae and conidia up to 10 μm in diameter

A. ochraceus (not described) has rough stalks and large heads, but the conidia are yellow-brown in colour

SEXUAL STATE

None known.

CLINICAL IMPORTANCE

It is a cause of invasive aspergillosis in immunocompromised patients. In non-compromised individuals it can act as a potent allergen or cause localised infections of the paranasal sinuses or other sites.

ASPERGILLUS FUMIGATUS

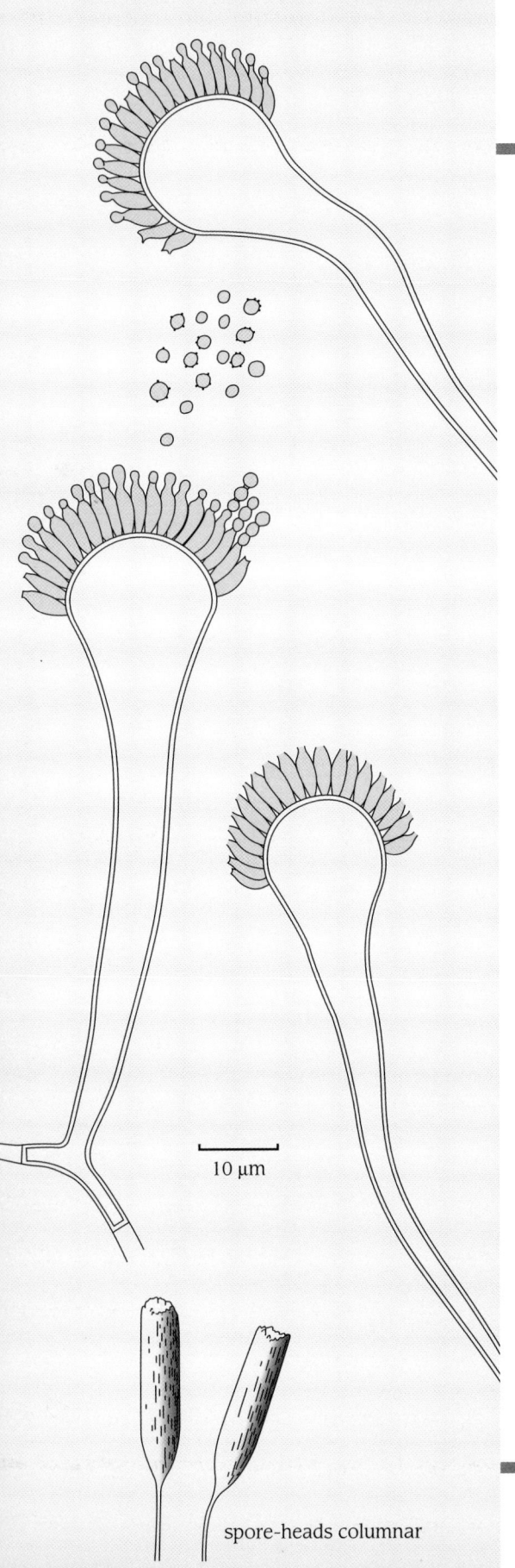

spore-heads columnar

COLONIAL APPEARANCE
at 30°C on glucose peptone agar

diameter	50 mm in one week
topography	flat, spreading
texture	powdery to felt-like
colour	blue-green, often with white margin
reverse	cream

MICROSCOPIC APPEARANCE
at 30°C

predominant features	vesiculate conidiophores; numerous small conidia, spore mass columnar
conidiophore	stalks short; vesicle pear-shaped; metulae absent; phialides crowded, pointing upwards, on upper two-thirds of vesicle only; some isolates have slightly brown-pigmented phialides
conidia	round, 2.5-3.5 µm, slightly roughened

VARIANT FORM

white form poorly sporulating colonies; often shows aberrant conidiogenous cell development

DIFFERENTIAL DIAGNOSIS

colonial appearance *Aspergillus clavatus* (not described), *A. nidulans, A. versicolor, Penicillium* spp.

microscopic appearance *A. glaucus,* but this species has conidiogenous cells over the entire surface of a round vesicle

SEXUAL STATE

None known.

CLINICAL IMPORTANCE

This common environmental organism is the predominant cause of aspergillosis in humans and animals. In non-compromised individuals it can act as a potent allergen or cause localised infection of the lungs, sinuses or other sites. In immunocompromised patients, inhalation of spores gives rise to invasive infection of the lungs or sinuses and dissemination to other organs often follows. This condition is lethal if left untreated.

ASPERGILLUS GLAUCUS

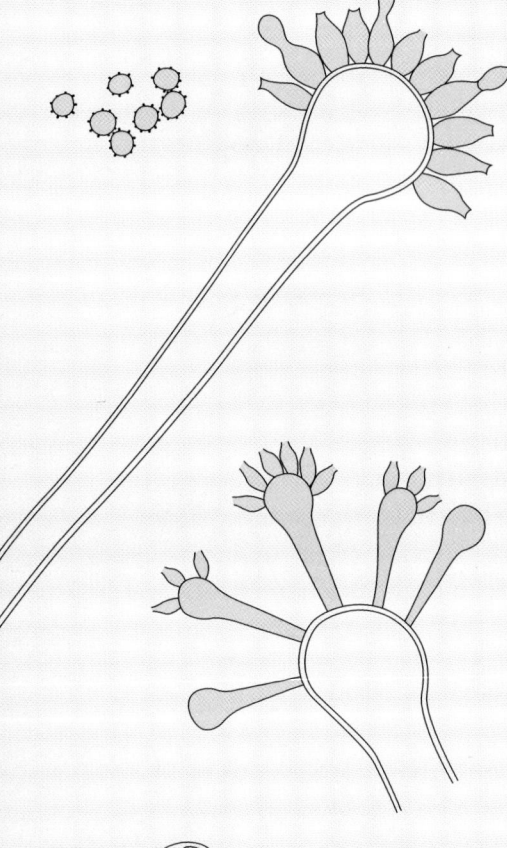

COLONIAL APPEARANCE
at 30°C on glucose peptone agar

diameter	20 mm in one week
topography	flat
texture	powdery to densely floccose
colour	pale blue-green becoming brownish green; yellow ascocarps often present
reverse	cream

MICROSCOPIC APPEARANCE
at 30°C

predominant features	vesiculate conidiophores; numerous small conidia; ascocarps often present
conidiophore	wide, thin-walled stalks; vesicles range in shape from club-shaped to round; metulae absent; phialides usually cover the entire surface of the vesicle; many isolates have aberrant heads with secondary conidiophores arising from the vesicle
conidia	round to oval, 4-8 µm, usually slightly roughened
ascocarps	non-pigmented, round, 80-250 µm (according to sexual state)

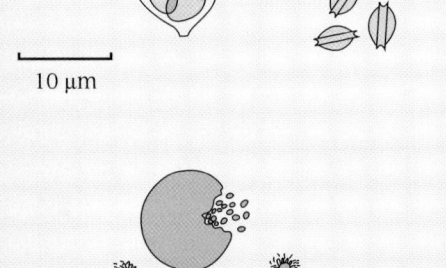

ascocarps (low power)

DIFFERENTIAL DIAGNOSIS

colonial appearance   *Aspergillus nidulans, A. versicolor, A. fumigatus, Penicillium* spp.

microscopic appearance   *A. fumigatus*

SEXUAL STATE

The *A. glaucus* group contains a number of similar species which are distinguished by the characteristics of their sexual states. These belong to the genus *Eurotium*.

CLINICAL IMPORTANCE

This common environmental organism is rarely a cause of human infection.

ASPERGILLUS NIDULANS

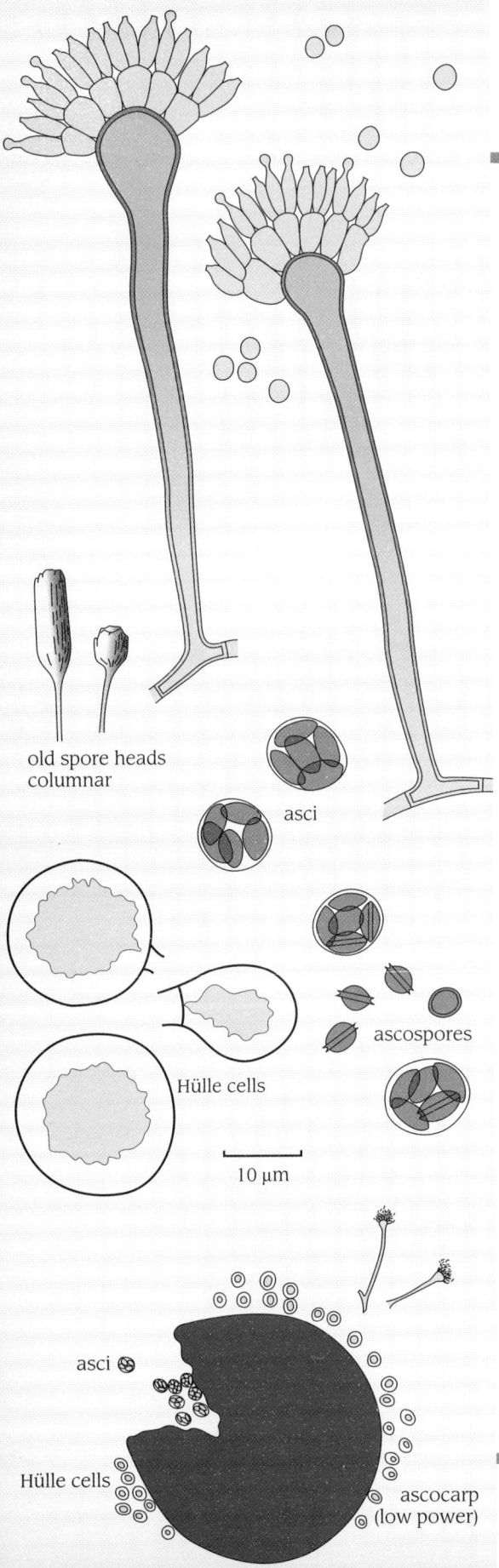

COLONIAL APPEARANCE
at 30°C on glucose peptone agar

diameter	30 mm in one week
topography	flat
texture	velvety to powdery
colour	dark green, developing yellow patches
reverse	deep red to purple

MICROSCOPIC APPEARANCE
at 30°C

predominant features	small, vesiculate conidiophore; large, round ascocarps with red contents; abundant, colourless Hülle cells
conidiophore	smooth, brown-pigmented stalks with distinct foot cells; hemispherical vesicles with metulae and phialides on upper half; cleistothecia within yellow patches of Hülle cells; red or purple ascospores, 5 μm × 4 μm, with two equatorial crests
conidia	round to oval, 3-3.5 μm, smooth

DIFFERENTIAL DIAGNOSIS

colonial appearance   *Aspergillus fumigatus, A. glaucus, A. versicolor; Penicillium* spp.

microscopic appearance   *A. versicolor*, but the conidiogenous cells are distributed all over the vesicles and the stalk is uncoloured

SEXUAL STATE

Emericella nidulans

CLINICAL IMPORTANCE

It is a cause of pale-grain mycetoma and rarely of deep infection in immunocompromised patients.

ASPERGILLUS VERSICOLOR

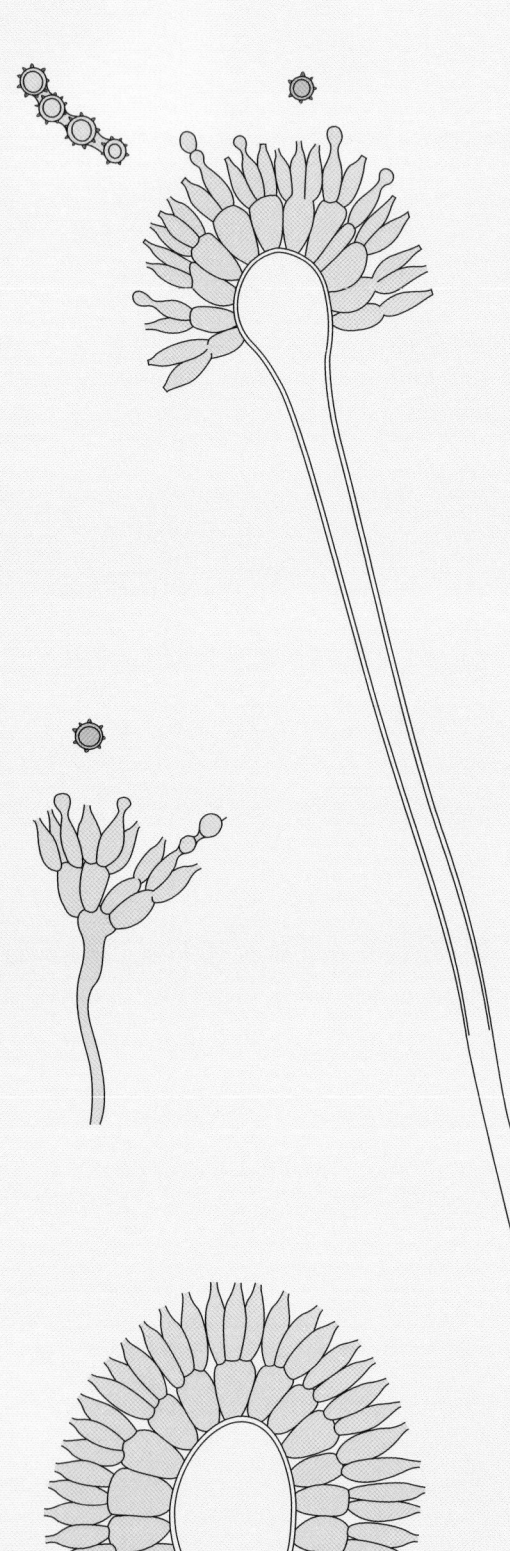

COLONIAL APPEARANCE
at 30°C on glucose peptone agar

diameter	20 mm in one week
topography	flat
texture	granular, floccose or velvety
colour	shades of green, yellow, light brown and pink; considerable variation in individual colonies
reverse	pale cream to red

MICROSCOPIC APPEARANCE
at 30°C

predominant features	vesiculate conidiophores with radiate heads
conidiophore	stalks smooth, colourless; vesicles round to oval with metulae and phialides over entire surface; reduced Penicillium-like heads also present; Hülle cells sometimes produced
conidia	round, 2.5-3 µm, slightly roughened

DIFFERENTIAL DIAGNOSIS

colonial appearance   *Aspergillus nidulans, Penicillium* spp.

microscopic appearance   *A. nidulans, A. ustus, Penicillium* spp.

SEXUAL STATE

None known.

CLINICAL IMPORTANCE

It is a rare cause of human disease, particularly nail infection.

ASPERGILLUS USTUS

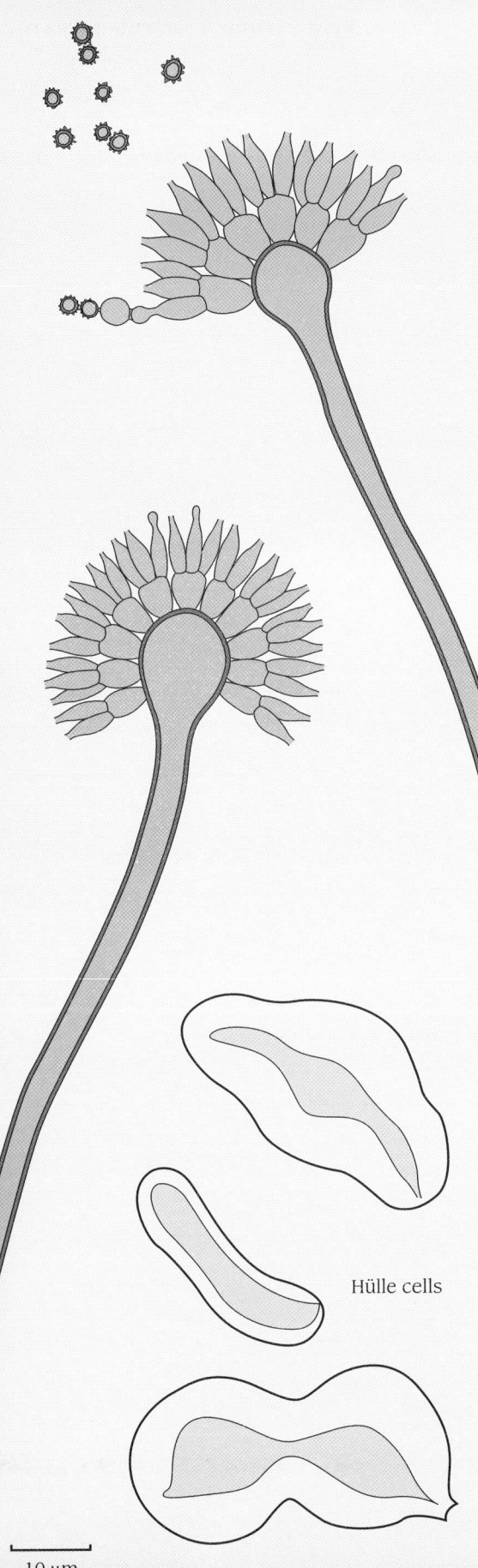

COLONIAL APPEARANCE
at 30°C on glucose peptone agar

diameter	30 mm in one week
topography	flat
texture	floccose
colour	brownish-yellow becoming purple-grey to grey
reverse	yellow, reddish or purple

MICROSCOPIC APPEARANCE
at 30°C

predominant features	pale brown, vesiculate conidiophores; dark, rough conidia; some strains have large, oval Hülle cells
conidiophore	short, pale brown stalks arising from prominent foot cells; round vesicles with metulae and phialides on upper two-thirds; oval or sausage-shaped Hülle cells, frequently bent and twisted, may be abundant
conidia	dark brown, round, 3-3.5 μm, roughened

DIFFERENTIAL DIAGNOSIS

colonial appearance   *Scedosporium apiospermum*

microscopic appearance   *Aspergillus nidulans, A. versicolor*

SEXUAL STATE

None known.

CLINICAL IMPORTANCE

It is a rare cause of human disease.

ASPERGILLUS NIGER

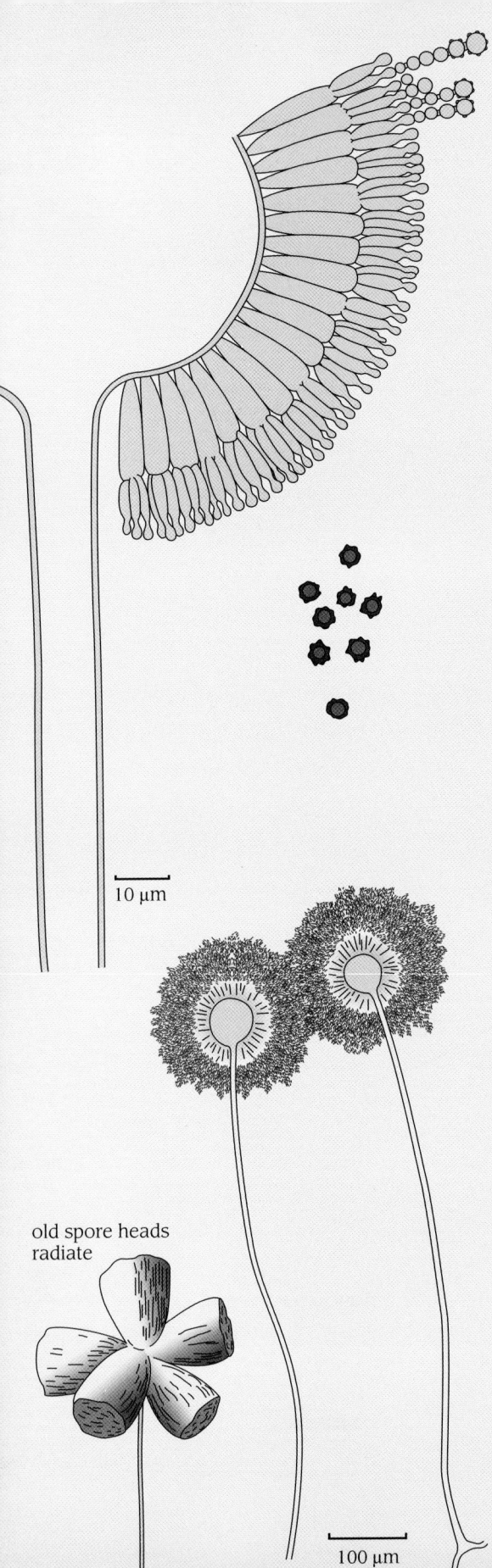

COLONIAL APPEARANCE
at 30°C on glucose peptone agar

diameter	60 mm in one week
topography	flat, often with radial folds
texture	granular
colour	white to yellow mycelium, developing a covering of black or purple-black sporing heads
reverse	cream

MICROSCOPIC APPEARANCE
at 30°C

predominant features	large, black sporing heads; black conidia
conidiophore	thick-walled, smooth, colourless stalks; large, round vesicle with phialides and metulae over entire surface
conidia	round to oval, 2.5-10 µm, roughened

DIFFERENTIAL DIAGNOSIS

colonial appearance *Aspergillus ustus*

microscopic appearance *A. candidus* has heads of similar structure, though the conidia are white

A. ochraceus (not described) produces similar large heads, but has rough stalks and pale brown conidia

SEXUAL STATE

None known.

CLINICAL IMPORTANCE

It is the most common cause of otomycosis in non-compromised individuals, and an occasional cause of deep infection in immunocompromised patients.

ASPERGILLUS TERREUS

COLONIAL APPEARANCE
at 30°C on glucose peptone agar

diameter	40 mm in one week
topography	flat
texture	granular to velvety
colour	cinnamon brown
reverse	yellow to pale brown

MICROSCOPIC APPEARANCE
at 30°C

predominant features	pale brown, columnar sporing heads; vesiculate conidiophores with characteristic fan-shaped heads
conidiophore	smooth, colourless stalks; dome-shaped vesicles with long, cylindrical metulae and phialides on upper two-thirds; thick-walled, oval, hyaline chlamydospores may be produced on the sides of the submerged vegetative hyphae
conidia	round, 2 µm in diameter, smooth

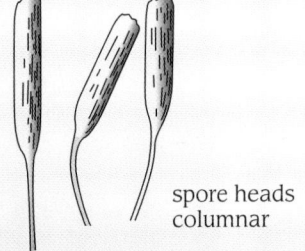

spore heads columnar

DIFFERENTIAL DIAGNOSIS

colonial appearance	*Scopulariopsis brevicaulis, Paecilomyces variotii, Myceliophthora thermophila*
microscopic appearance	other *Aspergillus* spp. with metulae, but the columnar appearance of the sporing heads and the small conidia should help to differentiate this species from the others

SEXUAL STATE

None known.

CLINICAL IMPORTANCE

It is a cause of onychomycosis and otomycosis in non-compromised individuals. It is also a rare cause of deep aspergillosis in immunocompromised patients.

ASPERGILLUS CANDIDUS

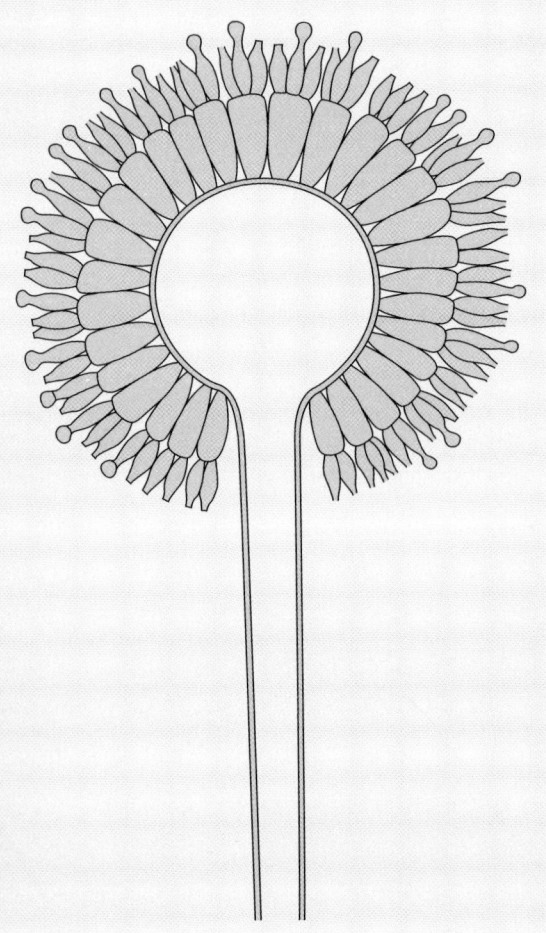

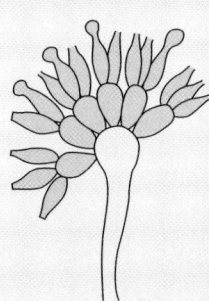

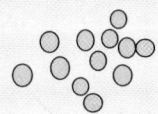

COLONIAL APPEARANCE
at 30°C on glucose peptone agar

diameter	20 mm in one week
topography	flat to domed
texture	granular to floccose
colour	white to pale cream
reverse	pale cream

MICROSCOPIC APPEARANCE
at 30°C

predominant features	large, white vesiculate conidiophores, mixed with non- or poorly vesiculate heads
conidiophore	larger heads have large, round vesicles with phialides and metulae covering entire surface, or occasionally on upper one-third only; smaller heads may lack vesicles and resemble Penicillium conidiophores
conidia	round to oval, colourless, 2.5-3.5 µm, smooth

DIFFERENTIAL DIAGNOSIS

colonial appearance   *Trichophyton terrestre*, some other floccose dermatophytes

microscopic appearance   *Aspergillus niger* and *A. ochraceus*, though the colourless conidia of *A. candidus* distinguish this species from the others

SEXUAL STATE

None known.

CLINICAL IMPORTANCE

It is a rare cause of onychomycosis and otomycosis.

PENICILIUM MARNEFFEI

(Hazard Group 3 pathogen)

COLONIAL APPEARANCE
at 30°C on glucose peptone agar

diameter	20 mm in one week
topography	flat to wrinkled
texture	glabrous, or felt-like
colour	light green to grey-green, with areas of pink
reverse	red with diffusing pigment

at 37°C

topography	yeast-like colonies
texture	glabrous
colour	cream to pale brown
reverse	cream

MICROSCOPIC APPEARANCE
at 30°C

predominant features	short, spreading conidiophores; small, lemon-shaped conidia
conidiophore	small heads on short stalks, with three to five divergent metulae (wider at the tip than the base), each producing broadly oval, tapering phialides
conidia	smooth, ellipsoidal, 2.5-4 µm × 2-3 µm, in long divergent chains with prominent disjunctors between the spores

at 37°C

predominant features	irregular hyphal fragments and cylindrical arthrospores

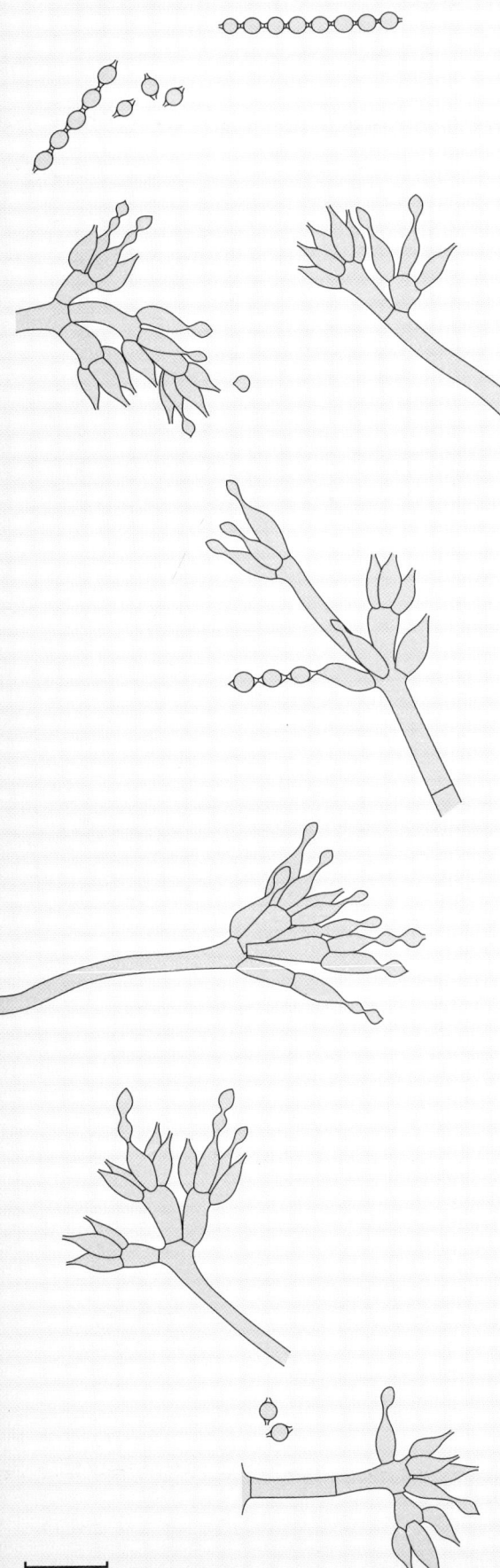

10 µm

DIFFERENTIAL DIAGNOSIS

colonial appearance

- **at 30°C**   other *Penicillium* spp., *Aspergillus* spp.

- **at 37°C**   *Trichosporon* spp., *Blastomyces dermatitidis*

microscopic appearance

- **at 30°C**   other *Penicillium* spp.

- **at 37°C**   *Trichosporon* spp.

SEXUAL STATE

None known.

CLINICAL IMPORTANCE

This unusual dimorphic fungus has a restricted geographical distribution and all natural human infections have occurred in individuals who resided in or visited the endemic regions of South East Asia. Disseminated infection has occurred in patients with AIDS following visits to Thailand. Penicilliosis is a treatable condition, but is often fatal if left untreated. The organism may pose a serious threat to laboratory workers handling live cultures.

SCOPULARIOPSIS BREVICAULIS

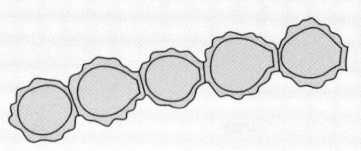

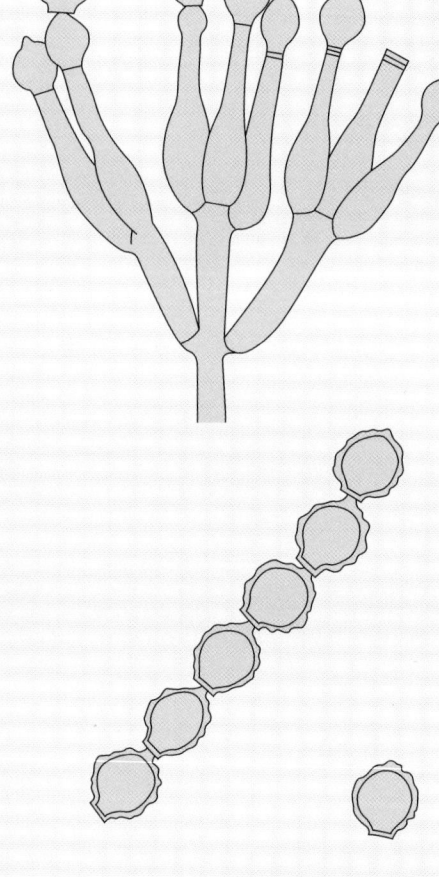

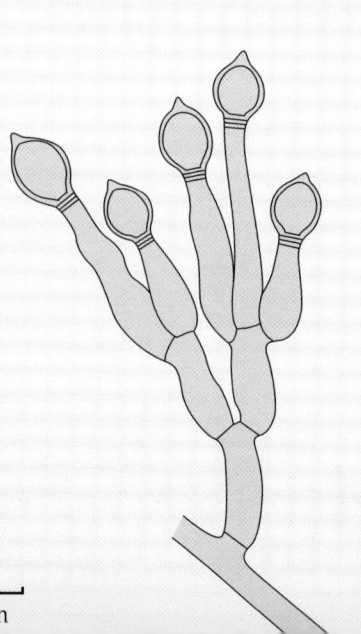

COLONIAL APPEARANCE
at 30°C on glucose peptone agar

diameter	50 mm in one week
topography	flat
texture	smooth, velvety to thickly granular, becoming loosely floccose
colour	pale to rich sand-brown
reverse	pale brown

MICROSCOPIC APPEARANCE
at 30°C

predominant features	chains of large, round, rough conidia; Penicillium-like conidiophores
conidiophore	short, branched heads, terminating in wide-necked, phialide-like cells which show annellations in older cultures
conidia	lemon-shaped, 6-7 μm long, with prominent flattened base, roughened

DIFFERENTIAL DIAGNOSIS

colonial appearance *Aspergillus terreus, Myceliophthora thermophila*

microscopic appearance other *Scopulariopsis* spp., but these have mostly white or dark grey colonies; *Penicillium* spp.; *Monascus ruber*

SEXUAL STATE

None known.

CLINICAL IMPORTANCE

It is a well recognised cause of nail infection in non-compromised individuals, and an occasional cause of deep infection in immunocompromised patients.

PAECILOMYCES LILACINUS

COLONIAL APPEARANCE
at 30°C on glucose peptone agar

diameter	30 mm in one week
topography	flat to domed
texture	densely floccose
colour	white, becoming lilac
reverse	pale or deep purple

MICROSCOPIC APPEARANCE
at 30°C

predominant features	chains of small conidia; Penicillium-like conidiophores
conidiophore	stalks sometimes roughened; irregularly branched heads terminating in long, tapering phialides; some single phialides borne along the sides of hyphae
conidia	ellipsoidal, 2.5-3 µm × 2 µm, smooth

DIFFERENTIAL DIAGNOSIS

colonial appearance *Fusarium* spp., *Monascus ruber*

microscopic appearance other *Paecilomyces* spp.; *Penicillium* spp.

SEXUAL STATE

None known.

CLINICAL IMPORTANCE

It has been implicated in several outbreaks of post-surgical endophthalmitis. It is a rare cause of deep infection in immunocompromised patients.

PAECILOMYCES VARIOTII

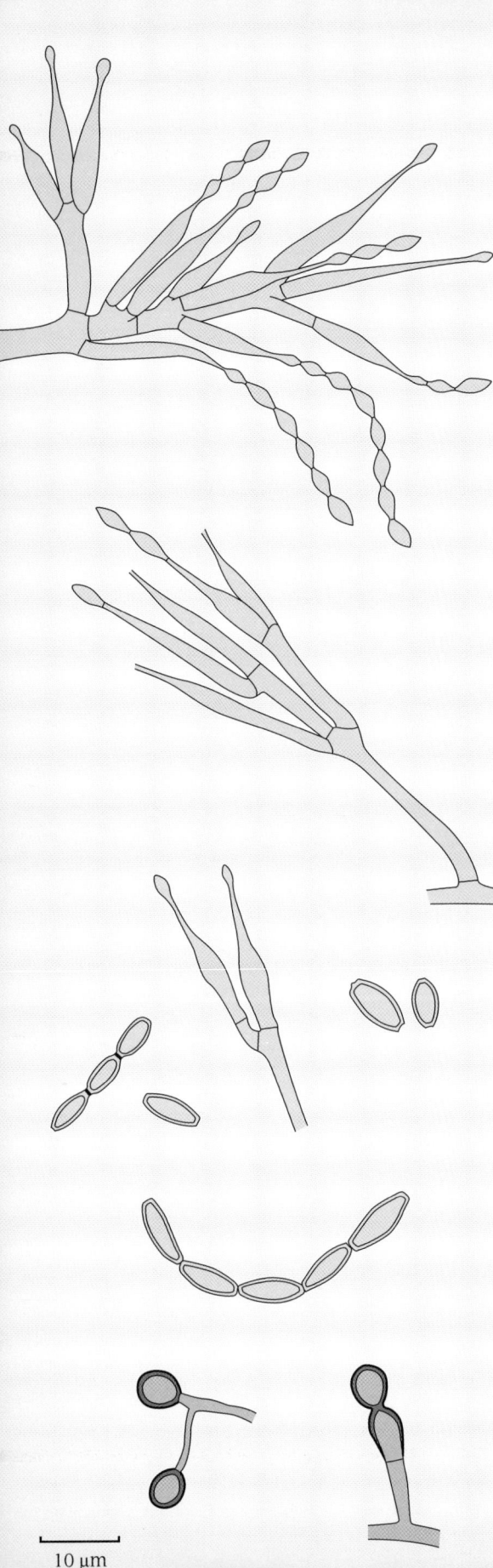

COLONIAL APPEARANCE
at 30°C on glucose peptone agar

diameter	50 mm in one week
topography	flat
texture	granular to loosely floccose
colour	olive-brown
reverse	pale cream

MICROSCOPIC APPEARANCE
at 30°C

predominant features	long chains of large, elliptical conidia; Penicillium-like conidiophores
conidiophore	irregularly branched heads terminating in long, tapering phialides; some single phialides borne along the sides of hyphae
conidia	ellipsoidal, 5-7 μm × 2.5-3 μm, smooth

VARIANT FORM

dark brown form   colonies dark brown due to abundant production of pigmented chlamydospores

DIFFERENTIAL DIAGNOSIS

colonial appearance   *Aspergillus terreus*, *Scopulariopsis brevicaulis*

- dark brown form   *Alternaria* spp., *Phialophora* spp., other dematiaceous fungi

microscopic appearance   other *Paecilomyces* spp., *Penicillium* spp.

SEXUAL STATE

None known.

CLINICAL IMPORTANCE

It is a rare cause of deep infection in immunocompromised patients.

8 MOULDS WITH ENTEROBLASTIC C

INTRODUCTION

The moulds described in this chapter are like those described in Chapter 7 in that they form enteroblastic conidia from phialidic or annellidic cells but in contrast they produce conidia that are adapted for water dispersal rather than in chains for aerial dissemination. These conidia are coated in a wettable slime and adhere to the tip of the conidiogenous cell in a mass. This feature is best seen by examining the colonies with a dissecting microscope because the slime is dispersed in wet preparations and the conidia float in the mounting fluid. The spore-bearing structures are often minute and are best studied using an oil immersion objective on a slide culture preparation.

The following descriptions include a number of moulds that form white, pink or red colonies. Those which produce multicellular macroconidia in culture can be identified as *Fusarium* or *Cylindrocarpon* species. However, many Fusarium isolates produce macroconidia late or in small numbers on common media, and the fungus presents with abundant microconidia resembling *Acremonium* species. In general, Acremonium

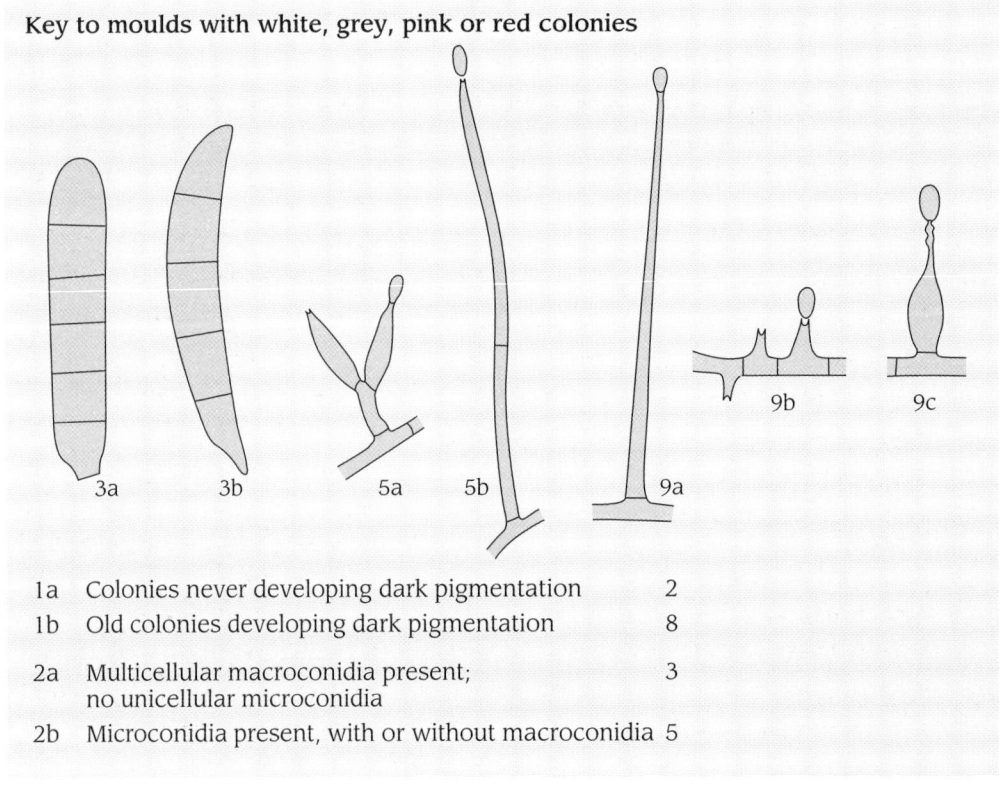

Key to moulds with white, grey, pink or red colonies

1a	Colonies never developing dark pigmentation	2
1b	Old colonies developing dark pigmentation	8
2a	Multicellular macroconidia present; no unicellular microconidia	3
2b	Microconidia present, with or without macroconidia	5

...IDIA ADHERING IN WET MASSES

has narrow, tapering phialides which appear straighter than those of Fusarium and these often arise from bundles of hyphae. In addition, the conidia of Fusarium are ovoid or kidney-shaped, and these tend to be larger than the more cylindrical Acremonium conidia. Formation of chains of conidia is a characteristic of *F. moniliforme*, but it is difficult to demonstrate except on media low in nutrients. On glucose peptone agar the conidia of *F. moniliforme* form into the globular masses seen with other species. Another feature of use in distinguishing some *Fusarium* species is the polyphialide, a phialidic cell with more than one conidiogenous site.

The genera *Lecythophora* and *Phialemonium* were created out of *Acremonium* to accommodate those species with so-called adelophialides, in which the phialidic sites are on lateral outgrowths of hyphal cells. Unlike typical phialides, these are not delimited from the hyphal cell by a septum.

The annellidic nature of the conidiogenous cells of *Scedosporium* species also presents some difficulties in identification, in that the minute annellidic scars at the cell tip are

3a	Macroconidia straight, with round ends	*Cylindrocarpon lichenicola*
3b	Macroconidia curved, with pointed ends	4
4a	Macroconidia mostly two-celled	*Fusarium dimerum*
4b	Macroconidia on polyphialides	*Fusarium semitectum*
5a	Microconidia on short phialides	6
5b	Microconidia on long phialides	7
6a	Microconidia elongated; some forming chains	*Fusarium moniliforme*
6b	Microconidia ovoid to kidney shaped; not in chains	*Fusarium oxysporum*
7a	Conidia mostly wider than 2 µm	*Fusarium solani*
7b	Conidia narrower than 2 µm	*Acremonium strictum*
8a	Colonies mostly glabrous, at least near the edge	9
8b	Colonies mostly floccose	10
9a	Long, tapering phialides	*Acremonium kiliense* or *Phialophora parasitica*
9b	Phialides reduced to short outgrowths of hyphal cells	*Lecythophora mutabilis*
9c	Short annellides with swollen bases	*Scedosporium prolificans*
10a	Conidia large, 6-12 µm long	*Scedosporium aspiospermum*
10b	Conidia smaller, 3-7 µm long	*Phialophora parasitica*

easily overlooked. Wet mounts usually show a single spore at the tip, but a slide culture will show the sequential production of the conidia. The conidia of this genus are similar in size and appearance to the chlamydospores of a number of other fungi (*Paecilomyces* species, *Lecythophora* species and *Acremonium kiliense*) and careful observation of their development is required to avoid confusion in identification.

The following descriptions include a number of melanised fungi that are black or dark olive-brown in colour, many of which have more than one growth form and the same isolate will often produce hyphae with conidia, yeast cells, and chains of pseudohyphal cells. These species are often termed the 'black yeasts'. In the *Phialophora* species the conidia are produced from phialides with definite collarettes. In the *Exophiala* species, the conidia are produced from annellides with irregular narrow peg-like tips. Both forms of conidiogenous cell are variants of the same basic process and *Exophiala* species occasionally produce well-formed collarettes. In addition, some isolates of *Phialophora verrucosa* produce holoblastic conidia of the Fonsecaea type.

Key to moulds with brown or black colonies

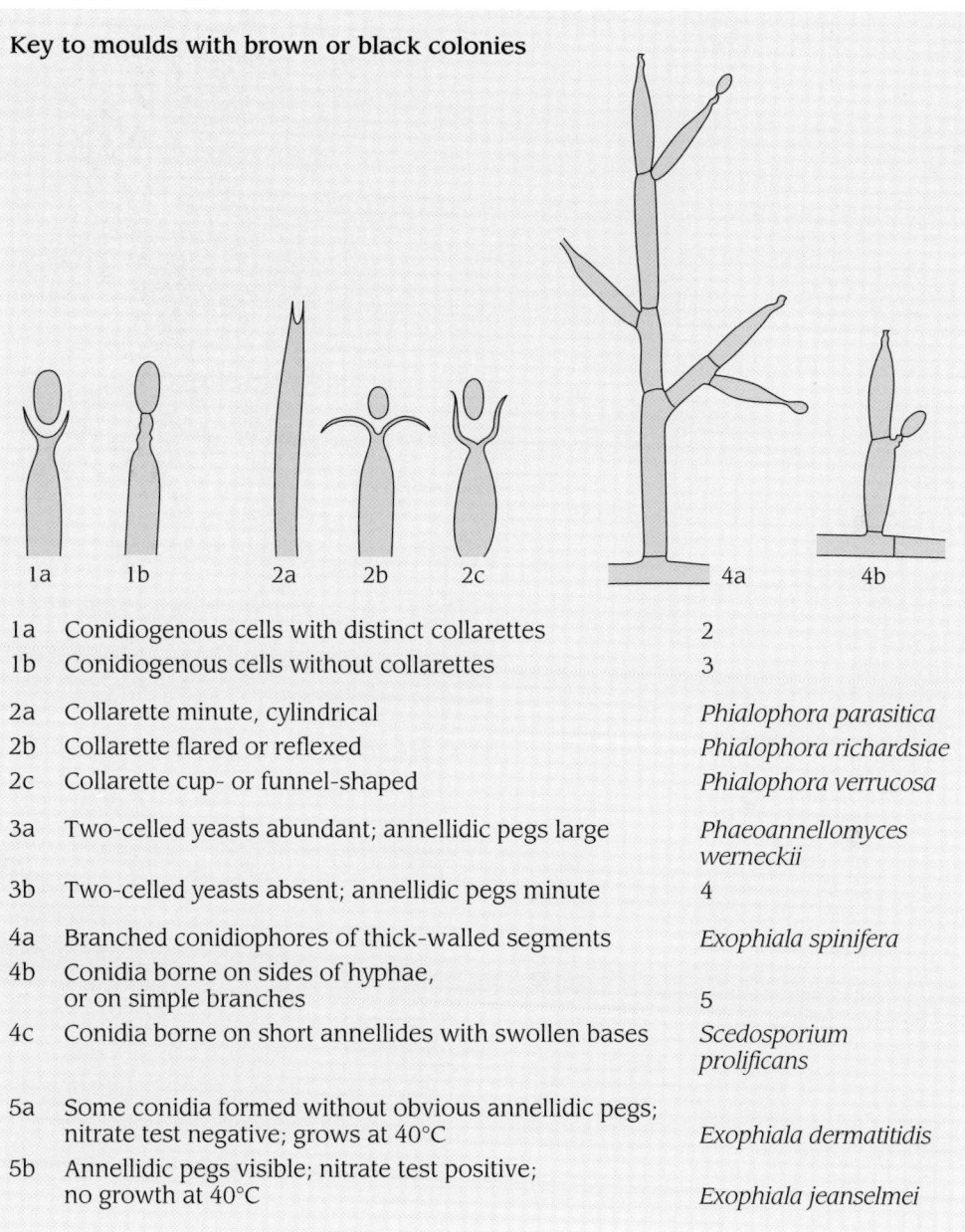

1a	Conidiogenous cells with distinct collarettes	2
1b	Conidiogenous cells without collarettes	3
2a	Collarette minute, cylindrical	*Phialophora parasitica*
2b	Collarette flared or reflexed	*Phialophora richardsiae*
2c	Collarette cup- or funnel-shaped	*Phialophora verrucosa*
3a	Two-celled yeasts abundant; annellidic pegs large	*Phaeoannellomyces werneckii*
3b	Two-celled yeasts absent; annellidic pegs minute	4
4a	Branched conidiophores of thick-walled segments	*Exophiala spinifera*
4b	Conidia borne on sides of hyphae, or on simple branches	5
4c	Conidia borne on short annellides with swollen bases	*Scedosporium prolificans*
5a	Some conidia formed without obvious annellidic pegs; nitrate test negative; grows at 40°C	*Exophiala dermatitidis*
5b	Annellidic pegs visible; nitrate test positive; no growth at 40°C	*Exophiala jeanselmei*

CYLINDROCARPON LICHENICOLA

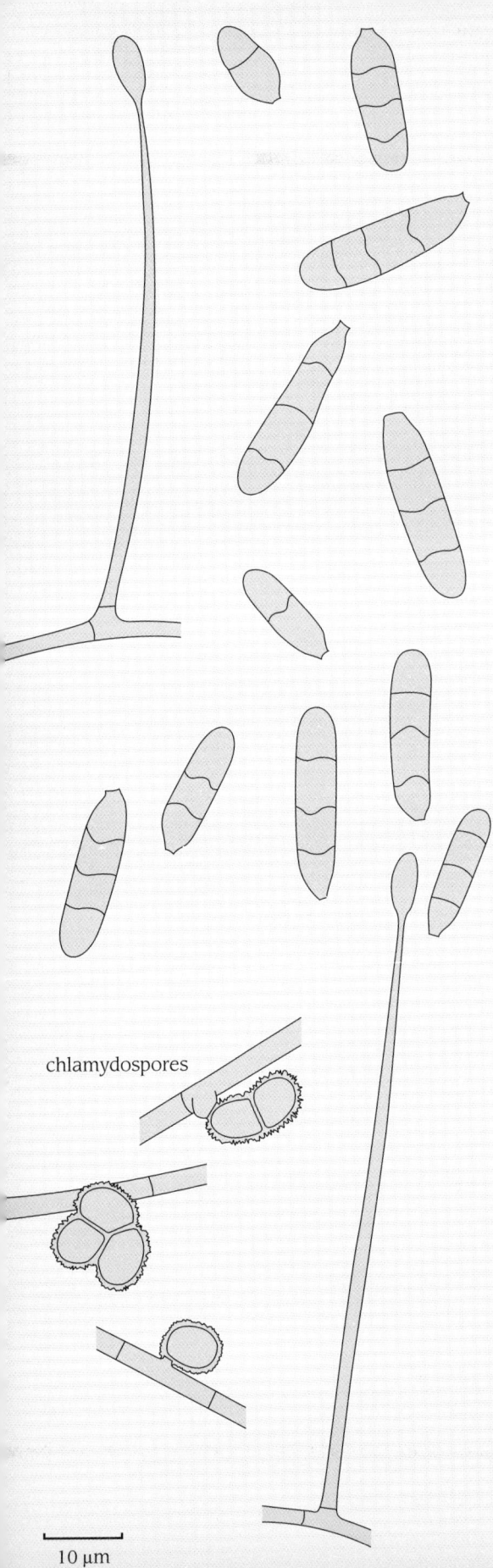

chlamydospores

10 μm

COLONIAL APPEARANCE
at 30°C on glucose peptone agar

diameter	50 mm in one week
topography	flat
texture	velvety to floccose
colour	pale brown to purple-red
reverse	dark brown

MICROSCOPIC APPEARANCE
at 30°C

predominant features	abundant, cylindrical conidia; in old cultures, abundant, round chlamydospores are formed singly in either chains or clusters
conidiophore	long, tapering phialides formed on usually simple but sometimes branched conidiophore
conidia	hyaline, cylindrical, 18-40 μm × 5-7 μm; sometimes slightly curved with three to five septa, rounded tip and truncate base

DIFFERENTIAL DIAGNOSIS

colonial appearance   other *Cylindrocarpon* spp.

microscopic appearance   other *Cylindrocarpon* spp., *Fusarium* spp. but the macroconidia of *Fusarium* spp. are curved, tapered to each end and possess a distinct foot cell

SEXUAL STATE

None known.

CLINICAL IMPORTANCE

It is an uncommon cause of human infection.

FUSARIUM DIMERUM

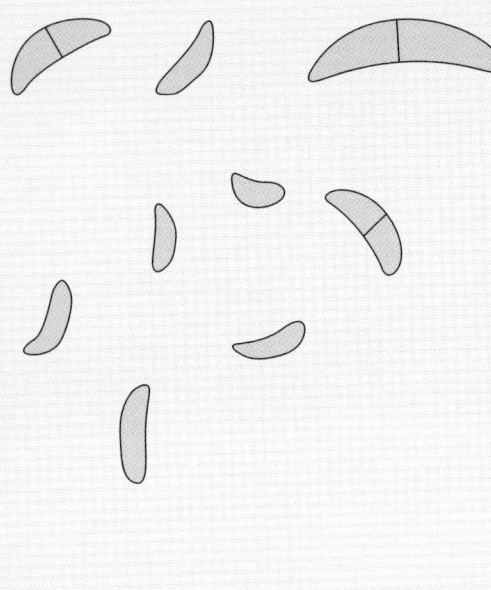

COLONIAL APPEARANCE
at 30°C on glucose peptone agar

diameter	30 mm in one week
topography	flat
texture	wet, sometimes floccose
colour	orange to apricot, aerial mycelium white
reverse	pale orange

MICROSCOPIC APPEARANCE
at 30°C

predominant features	crescent-shaped, two-celled macroconidia formed from short phialides; round chlamydospores may be present singly or in short chains
conidiophore	short, tapering phialides produced singly or in pairs
microconidia	absent, but there may be immature (unicellular) macroconidia
macroconidia	crescent-shaped, 5-25 µm × 1.5-4.2 µm; pointed at one end with one to three septa

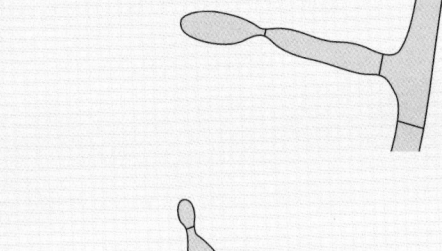

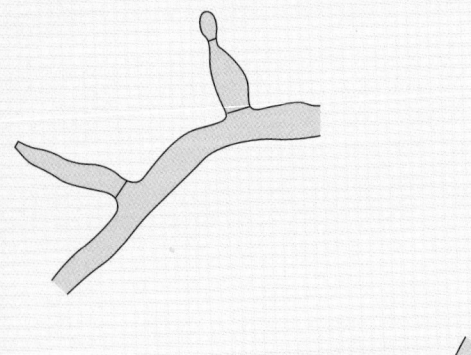

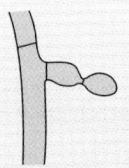

10 µm

DIFFERENTIAL DIAGNOSIS

colonial appearance *Acremonium* spp.

microscopic appearance other *Fusarium* spp.

SEXUAL STATE

None known.

CLINICAL IMPORTANCE

It has caused corneal infection and is a rare cause of disseminated infection in neutropenic patients.

FUSARIUM SEMITECTUM

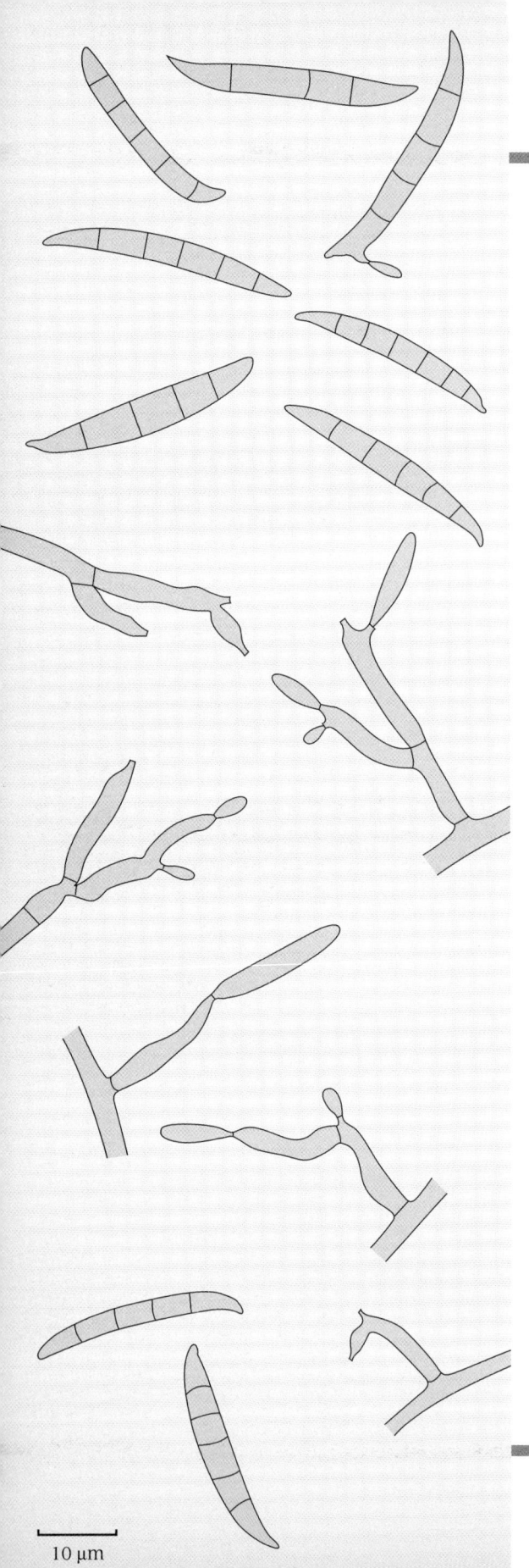

COLONIAL APPEARANCE
at 30°C on glucose peptone agar

diameter	60 mm in one week
topography	flat
texture	floccose
colour	white to pale apricot
reverse	peach or brown

MICROSCOPIC APPEARANCE
at 30°C

predominant features	abundant microconidia produced by polyphialides; no distinct division between microconidia and macroconidia
conidiophore	short, cylindrical phialides produced on short, branching conidiophores; initially with a single apical collarette, later developing into polyphialides with more than one collarette
microconidia	absent, but there may be some smaller 17-28 µm × 2.5-4 µm macroconidia with up to two septa
macroconidia	straight or slightly curved, 22-40 µm × 3-4.5 µm, with three to five septa; wedge-shaped foot cell

10 µm

DIFFERENTIAL DIAGNOSIS

colonial appearance   other *Fusarium* spp.

microscopic appearance   other *Fusarium* spp., *Acremonium* spp.

SEXUAL STATE

None known.

CLINICAL IMPORTANCE

It is a rare cause of human disease.

FUSARIUM MONILIFORME

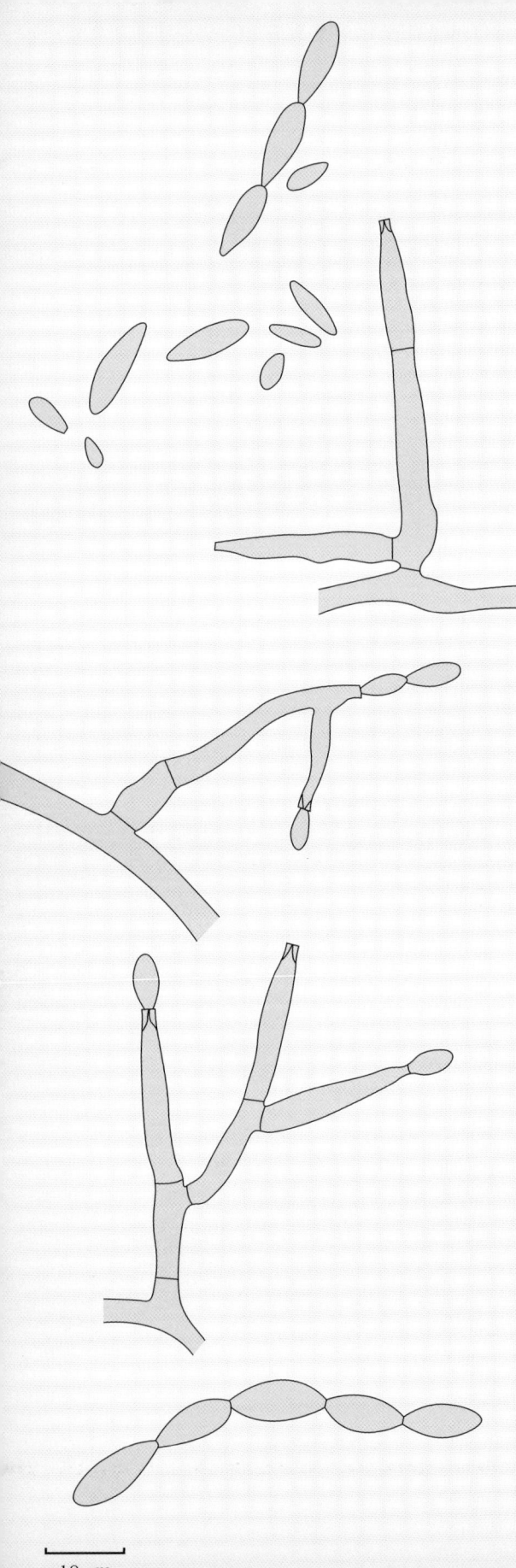

COLONIAL APPEARANCE
at 30°C on glucose peptone agar

diameter	40 mm in one week
topography	flat
texture	floccose
colour	white to peach or salmon pink becoming tinged with purple
reverse	pale cream

MICROSCOPIC APPEARANCE
at 30°C

predominant features	abundant, oval to club-shaped microconidia, sometimes formed in chains; there may be macroconidia
conidiophore	short, tapering phialides with inconspicuous collarettes; polyphialides with more than one collarette are common and may be arranged singly or in branching systems
microconidia	oval to club-shaped, 7-10 µm × 2.5-3.2 µm; borne in chains on some media (including cornmeal agar)
macroconidia	three to seven septa; 31-58 µm × 2.7-3.6 µm; ends may be slightly curved

DIFFERENTIAL DIAGNOSIS

colonial appearance other *Fusarium* spp., *Acremonium* spp.

microscopic appearance other *Fusarium* spp., but these seldom form microconidial chains; *Paecilomyces* spp.

SEXUAL STATE

None known.

CLINICAL IMPORTANCE

It is a cause of corneal infection in non-compromised patients and has caused lethal disseminated infection in neutropenic individuals.

FUSARIUM OXYSPORUM

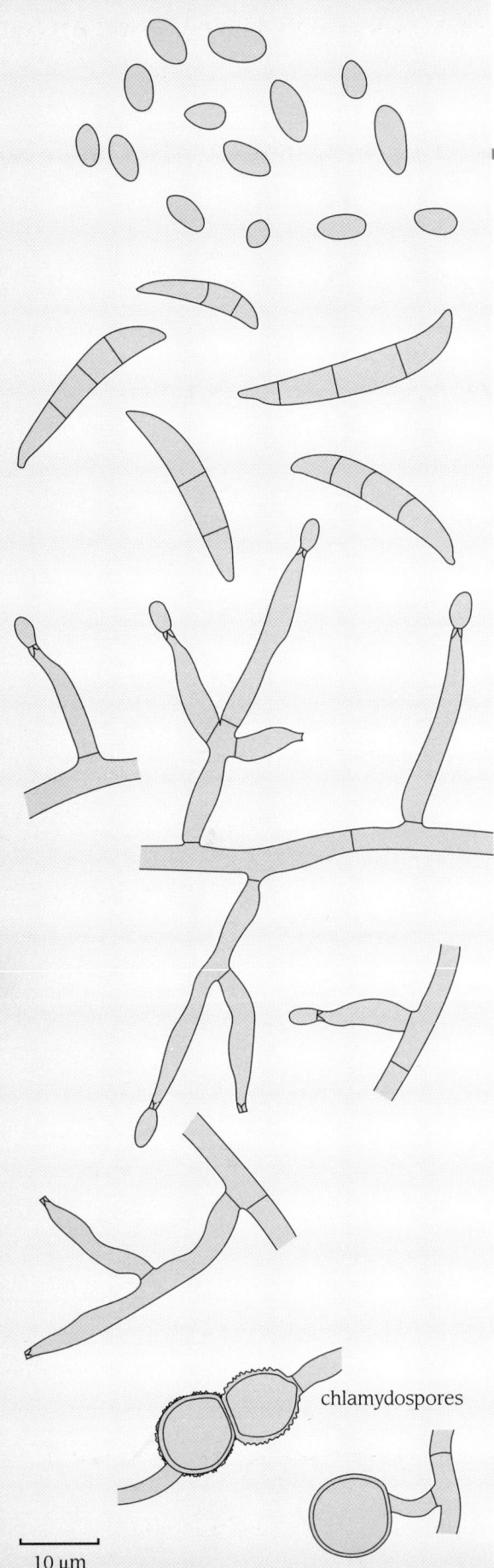

COLONIAL APPEARANCE
at 30°C on glucose peptone agar

diameter	50 mm in one week
topography	flat
texture	floccose becoming felted
colour	white to pale apricot, usually with a purple tinge
reverse	purple

MICROSCOPIC APPEARANCE
at 30°C

predominant features	abundant, small, oval microconidia mixed with smaller numbers of crescent-shaped macroconidia; large, round chlamydospores may be present, singly or in pairs
conidiophore	short, tapering phialides with inconspicuous collarettes; arranged singly or as branching conidiophores
microconidia	abundant, small, oval to kidney-shaped; occasionally with one or two septa
macroconidia	one to five septa, crescent-shaped with a definite foot cell and pointed distal end

DIFFERENTIAL DIAGNOSIS

colonial appearance   other *Fusarium* spp., *Acremonium* spp., *Cylindrocarpon* spp., *Paecilomyces lilacinus*

microscopic appearance   *F. solani*, but *F. oxysporum* may be distinguished by its shorter microconidial phialides

                              F. moniliforme, but this has polyphialides and can form chains of microconidia

SEXUAL STATE

None known.

CLINICAL IMPORTANCE

It is a common plant pathogen that can cause corneal and nail infection in non-compromised patients and lethal disseminated infection in neutropenic individuals.

FUSARIUM SOLANI

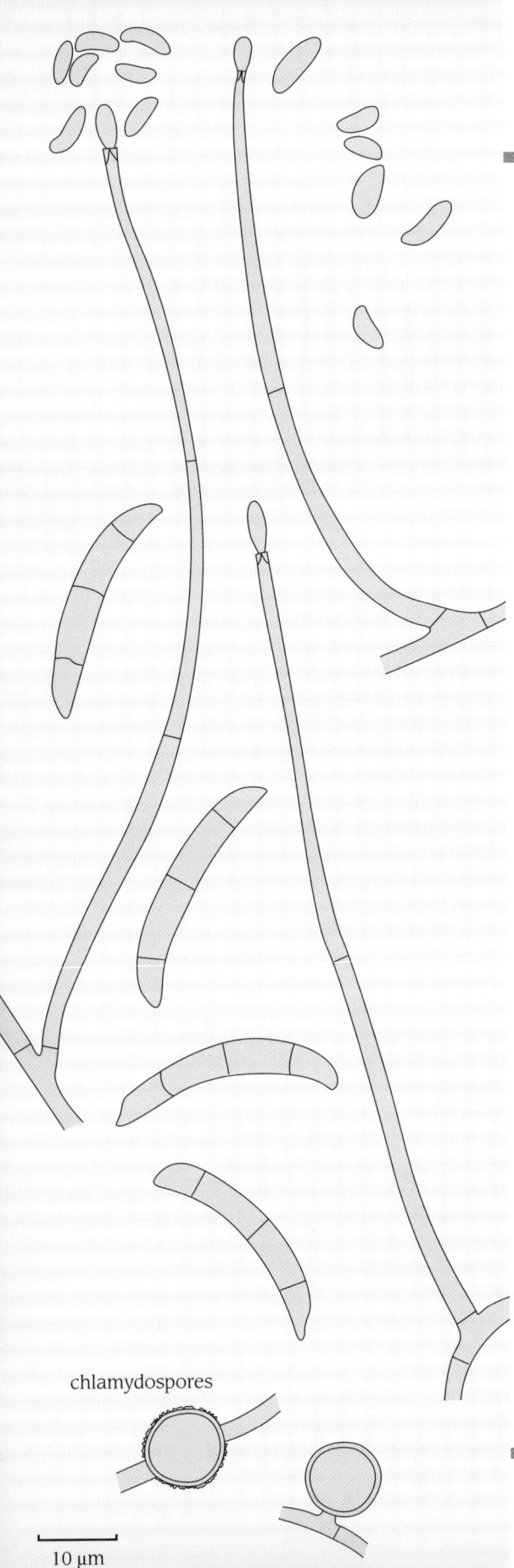

chlamydospores

10 µm

COLONIAL APPEARANCE
at 30°C on glucose peptone agar

diameter	30 mm in one week
topography	flat
texture	floccose
colour	greyish-white, cream, buff or pinkish-purple
reverse	pale cream

MICROSCOPIC APPEARANCE
at 30°C

predominant features	abundant, small, oval microconidia mixed with smaller numbers of crescent-shaped macroconidia; large, round chlamydospores may be present singly or in pairs
conidiophore	long, tapering phialides with inconspicuous collarettes; difficult to distinguish from vegetative hyphae
microconidia	abundant, small, oval to kidney-shaped; occasionally with one septum
macroconidia	one to five septa; crescent-shaped with a definite foot cell at one end

DIFFERENTIAL DIAGNOSIS

colonial appearance	other *Fusarium* spp., *Acremonium* spp., *Cylindrocarpon* spp., *Paecilomyces lilacinus*
microscopic appearance	*F. oxysporum*, but *F. solani* may be distinguished by its long, microconidial phialides

SEXUAL STATE

None known.

CLINICAL IMPORTANCE

It is a cause of nail, corneal and localised deep infection in non-compromised patients and has caused lethal disseminated infection in neutropenic individuals.

ACREMONIUM STRICTUM

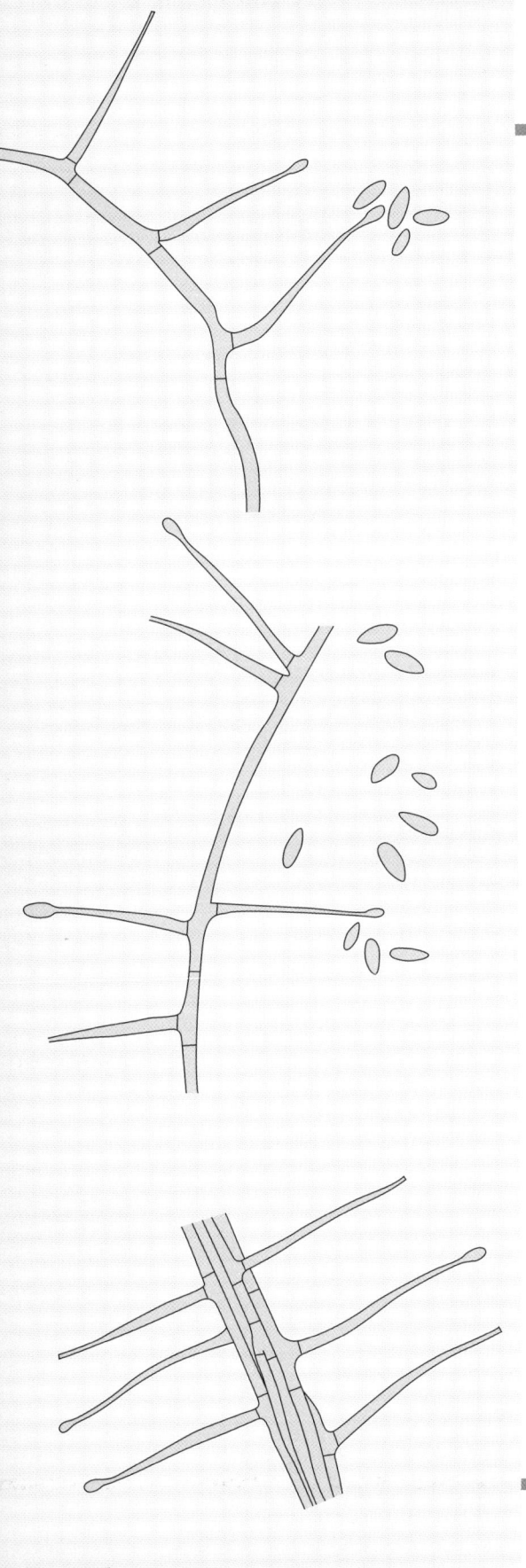

COLONIAL APPEARANCE
at 30°C on glucose peptone agar

diameter	50 mm in one week
topography	flat
texture	smooth, wet or velvety to floccose
colour	pink to orange
reverse	colourless or pale pink

MICROSCOPIC APPEARANCE
at 30°C

predominant features	long phialides arising from distinct bundles of aerial mycelium; balls of conidia accumulating at tips
conidiophore	long, slender phialides
conidia	cylindrical or ellipsoidal, 3.3–5.5 µm × 0.9–1.8 µm; formed in slimy balls at the tips of phialides

DIFFERENTIAL DIAGNOSIS

colonial appearance　　other *Acremonium* spp., but *A. strictum* has no diffusing pigment; *Fusarium* spp.

microscopic appearance　　other *Acremonium* spp., *Fusarium* spp.

SEXUAL STATE

None known.

CLINICAL IMPORTANCE

It is a rare cause of human disease.

ACREMONIUM KILIENSE

COLONIAL APPEARANCE
at 30°C on glucose peptone agar

diameter	50 mm in one week
topography	flat
texture	smooth
colour	grey to orange
reverse	brown with diffusing, dark pigment

MICROSCOPIC APPEARANCE
at 30°C

predominant features	balls of ellipsoidal conidia accumulated at the ends of long, slender phialides; oval chlamydospores
conidiophore	long, straight, slightly tapering phialides, arising as side-branches on hyphae
conidia	ellipsoidal, 3-6 μm × 1.5 μm, accumulating in slimy balls at the ends of phialides

DIFFERENTIAL DIAGNOSIS

colonial appearance   other *Acremonium* spp., *Fusarium* spp.

microscopic appearance other *Acremonium* spp., *Fusarium* spp.

SEXUAL STATE

None known.

CLINICAL IMPORTANCE

It is a cause of pale-grain mycetoma and rarely of deep infection in immunocompromised patients.

LECYTHOPHORA MUTABILIS

COLONIAL APPEARANCE
at 30°C on glucose peptone agar

diameter	20 mm in one week
topography	flat, becoming folded; with erect fascicles
texture	glabrous and yeast-like, becoming leathery with age
colour	white or pink at first, becoming brown-black with age
reverse	cream to pale pink, becoming brown with age

MICROSCOPIC APPEARANCE
at 30°C

predominant features	fasciculated hyphae; hyaline, oval conidia emerging from conidiogenous projections on the sides of hyphae; brown, thick-walled chlamydospores
conidiophore	small, cylindrical phialides tapering at apex with a collarette; not separated by a septum from the adjacent hypha
conidia	hyaline, oval to ellipsoidal, 4-6 μm × 1.8-2.5 μm; may be slightly kidney-shaped

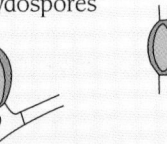

chlamydospores

10 μm

DIFFERENTIAL DIAGNOSIS

colonial appearance *Lecythophora hoffmannii, Sporothrix schenckii, Scedosporium prolificans, Aureobasidium pullulans, Trichosporon* spp.

microscopic appearance *L. hoffmannii, Phialophora* spp., *Phialemonium* spp.

SEXUAL STATE

None known.

CLINICAL IMPORTANCE

It is a rare cause of phaeohyphomycosis.

LECYTHOPHORA HOFFMANNII

This fungus differs from *L. mutabilis* by lacking chlamydospores and by possessing smaller 3.5 µm × 1.5 µm conidia. It is a rare cause of deep phaeohyphomycosis.

PHIALEMONIUM SPP.

Species of this genus are very similar to Lecythophora. However, the adelophialides are longer and more cylindrical and the collarette, which is less than 1.2 µm wide, is less obvious. These species are rare causes of human disease.

SCEDOSPORIUM PROLIFICANS

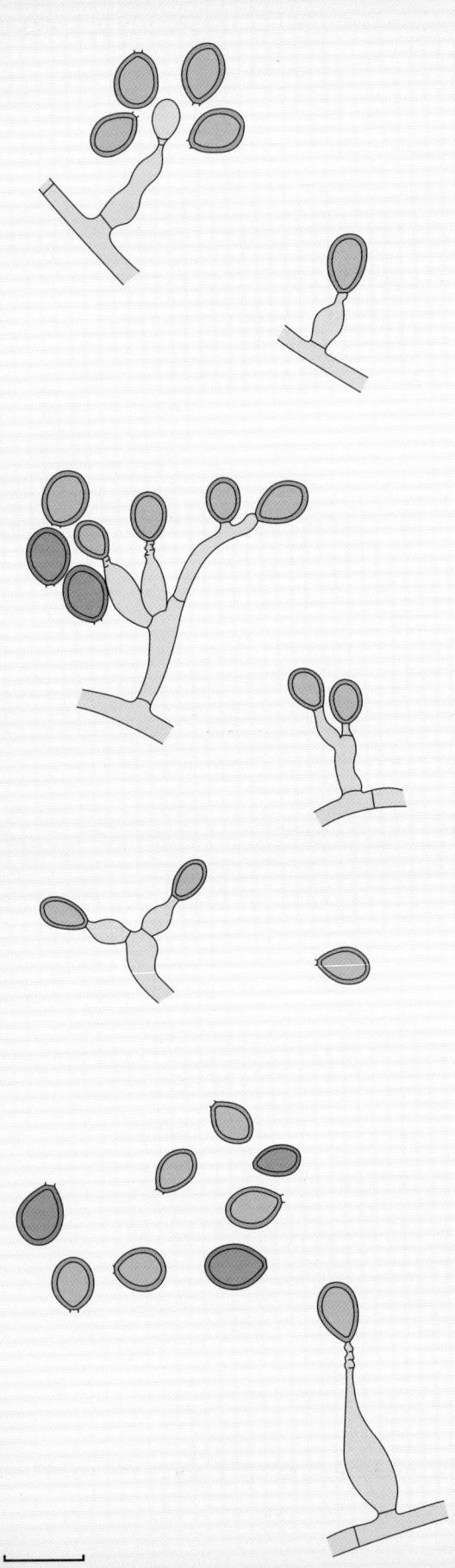

COLONIAL APPEARANCE
at 30°C on glucose peptone agar

diameter	20 mm in one week
topography	flat
texture	moist, with some floccose areas
colour	grey to black
reverse	black

MICROSCOPIC APPEARANCE
at 30°C

predominant features	numerous, dark brown, oval conidia formed from short annellides with inflated bases and narrow, tapering tips
conidiophore	short annellides with inflated bases, sometimes with long, tapering tips; arranged singly or in branched clusters
conidia	round to oval, dark brown, 3-7 µm × 2.5 µm; often appear to be in groups around the annellide tip

DIFFERENTIAL DIAGNOSIS

colonial appearance	*Aureobasidium pullulans, Lecythophora* spp., *Sporothrix schenckii, Exophiala* spp.
microscopic appearance	*Scedosporium apiospermum, Arthrinium* spp. (not described)

SEXUAL STATE

None known.

CLINICAL IMPORTANCE

It can cause serious deep-seated infection in immunocompromised individuals.

SCEDOSPORIUM APIOSPERMUM

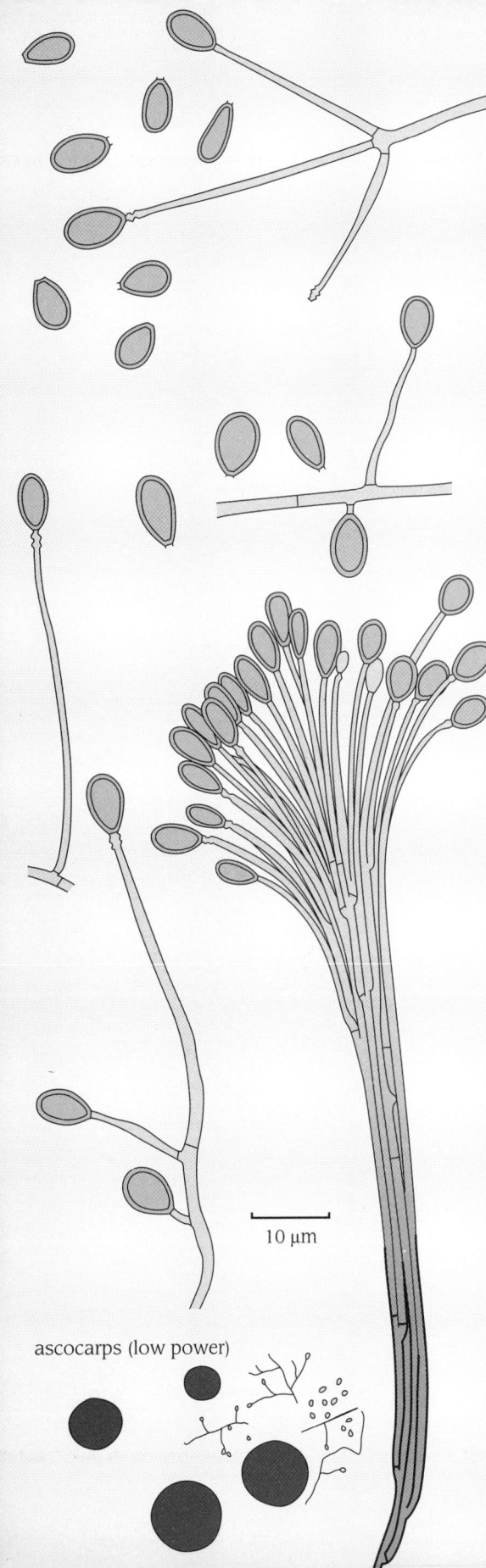

ascocarps (low power)

COLONIAL APPEARANCE
at 30°C on glucose peptone agar

diameter	40 mm in one week
topography	high, flat to dome-shaped
texture	floccose
colour	white to pale or dark grey
reverse	pale-yellow to dark-brown or black with age

MICROSCOPIC APPEARANCE
at 30°C

predominant features	abundant, oval conidia formed from single or branched annellides and sometimes aggregated into bundles; ascocarps sometimes present
conidiophore	long, slender annellides that are sometimes aggregated into bundles of tree-like synnemata (*Graphium* state); in slide preparations the annellidic tip usually has only a single conidium remaining attached
conidia	yellow to pale brown, oval, 6-12 µm × 3.5-6 µm, with a scar at the base; readily separating from the spore mass at the annellidic tip
ascocarps	black, round, with lemon-shaped ascospores

DIFFERENTIAL DIAGNOSIS

colonial appearance	*Aspergillus ustus*
microscopic appearance	other *Scedosporium* spp.; *Arthrinium* spp. (not described, but its colony is pure white and conidia black)

SEXUAL STATE

Pseudallescheria boydii

CLINICAL IMPORTANCE

This environmental fungus is the most common cause of pale-grain mycetoma in temperate regions. It is a well-recognised cause of pneumonia following aspiration of contaminated water, but it can also cause localised infection of other deep organs, as well as disseminated infection in immunocompromised patients. It may be found as a coloniser in the lungs of patients with cystic fibrosis.

PHIALOPHORA PARASITICA

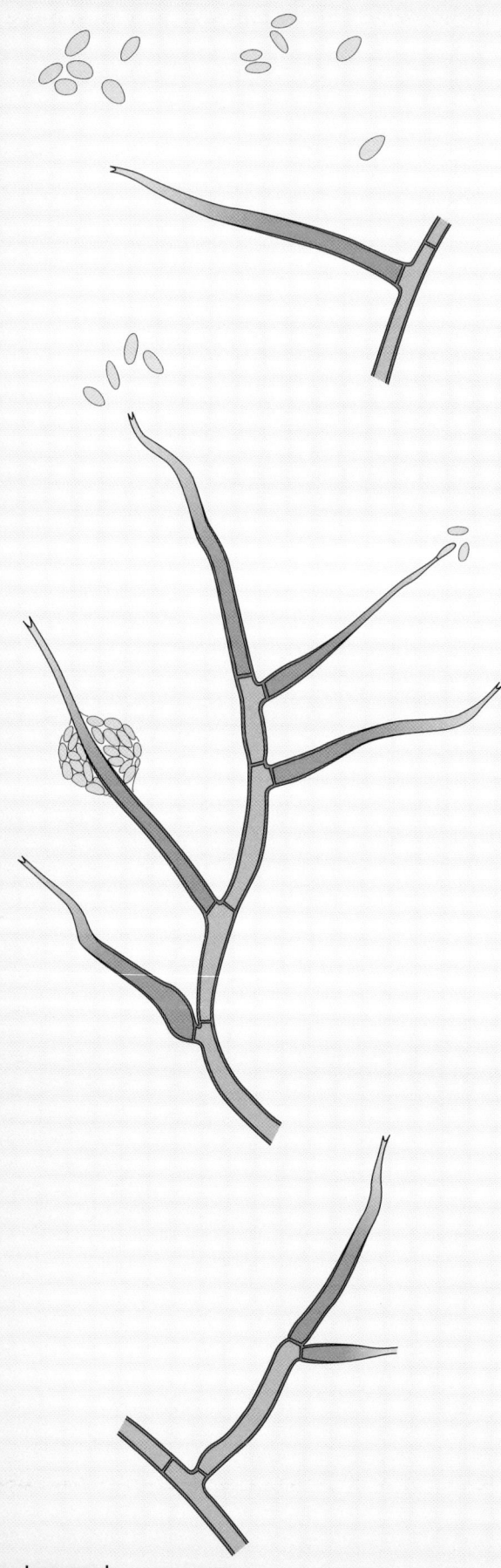

COLONIAL APPEARANCE
at 30°C on glucose peptone agar

diameter	40 mm in one week
topography	flat, may be folded
texture	glabrous to velvety with floccose centre
colour	white with grey-brown centre; becoming dark grey-brown with age
reverse	dark brown

MICROSCOPIC APPEARANCE
at 30°C

predominant features	thick-walled, brown phialides with inconspicuous ollarettes; small, oval conidia
conidiophore	thick-walled, brown, cylindrical phialides tapering towards tip with small, funnel-shaped collarette; secondary phialide often proliferating through the first
conidia	thin-walled, hyaline to pale brown, 3-6 μm × 1-2 μm, oval to slightly kidney-shaped

DIFFERENTIAL DIAGNOSIS

colonial appearance *Sporothrix schenckii, Scedosporium* spp., *Lecythophora* spp.

microscopic appearance *Phialophora richardsiae*, but the collarette of *P. parasitica* is never flared or reflexed

Acremonium spp., but conidiophores of *P. parasitica* become pigmented with age

SEXUAL STATE

None known.

CLINICAL IMPORTANCE

It is a cause of subcutaneous or deep phaeohyphomycosis and rarely of disseminated infection in immunocompromised individuals.

PHIALOPHORA RICHARDSIAE

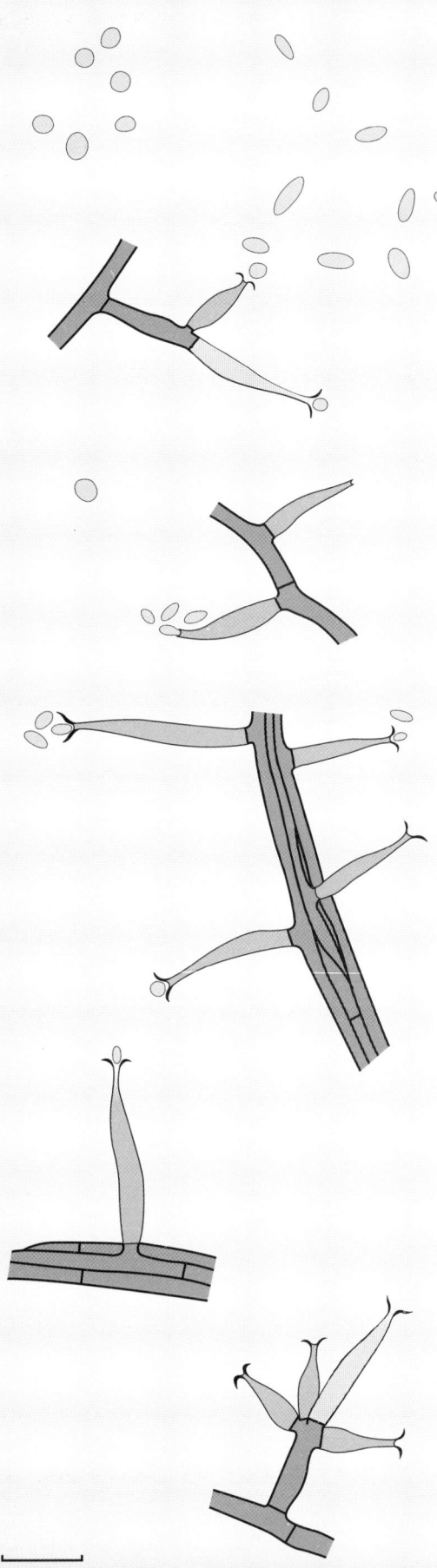

COLONIAL APPEARANCE
at 30°C on glucose peptone agar

diameter	25 mm in one week
topography	flat to domed
texture	velvety with floccose centre
colour	grey-brown to olive-brown
reverse	dark brown

MICROSCOPIC APPEARANCE
at 30°C

predominant features	thick-walled, brown phialides with flaring collarettes; oval, colourless conidia and round, brown conidia
conidiophore	thick-walled, brown, cylindrical phialides tapering towards tip, with widely flared or reflexed collarette
conidia	thin-walled, colourless, oval conidia, 2-4 µm × 1-2 µm; thick-walled, brown conidia 3 µm in diameter

DIFFERENTIAL DIAGNOSIS

colonial appearance   other *Phialophora* spp., *Exophiala* spp., *Fonsecaea* spp., *Cladosporium* spp.

microscopic appearance   *P. parasitica,* but the collarette is more developed in *P. richardsiae*

P. verrucosa, which has a more cup-shaped collarette

SEXUAL STATE

None known.

CLINICAL IMPORTANCE

It is a rare cause of subcutaneous phaeohyphomycosis.

PHIALOPHORA VERRUCOSA

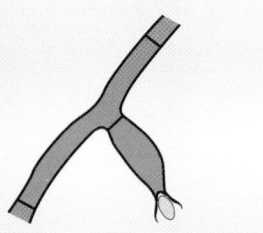
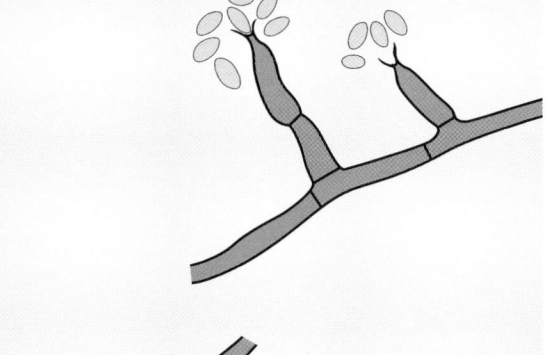
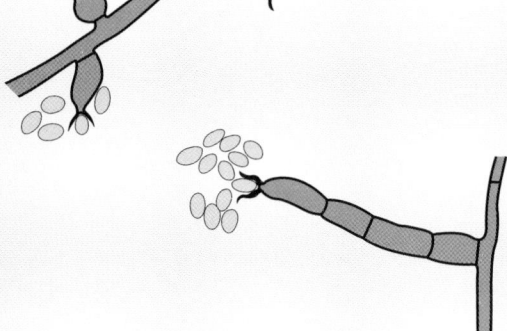

COLONIAL APPEARANCE
at 30°C on glucose peptone agar

diameter	10 mm in one week
topography	heaped, becoming flatter with age; may be folded
texture	hard and leathery, becoming velvety or floccose
colour	dark olive-grey to black
reverse	olive-grey

MICROSCOPIC APPEARANCE
at 30°C

predominant features	brown, septate hyphae; small phialides with cup-shaped collarettes; small, ellipsoidal conidia
conidiophore	brown, flask-shaped phialides produced from hyphae or on side branches with very distinct, cup-shaped, darkly pigmented collarettes that increase in size as successive conidia are produced
conidia	small, pale, ellipsoidal, 2.5-4 µm × 1.5-3 µm; accumulate in slimy clusters

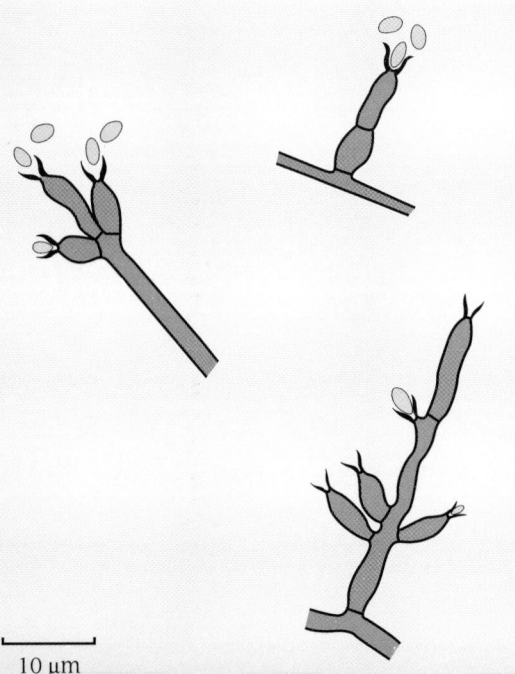

10 µm

DIFFERENTIAL DIAGNOSIS

colonial appearance other *Phialophora* spp., *Exophiala* spp., *Fonsecaea* spp., *Cladosporium* spp.

microscopic appearance other *Phialophora* spp., but *P. verrucosa* has a darkly pigmented, cup-shaped collarette

SEXUAL STATE

None known.

CLINICAL IMPORTANCE

It is one of the major causal organisms of chromoblastomycosis and rarely a cause of cerebral or disseminated infection in immunocompromised individuals.

PHAEOANNELLOMYCES WERNECKII

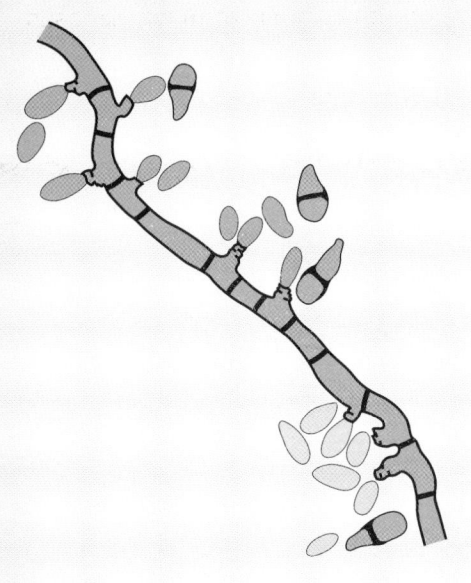

COLONIAL APPEARANCE
at 30°C on glucose peptone agar

diameter	15 mm in one week
topography	flat, becoming folded in centre
texture	glabrous; sometimes with floccose aerial mycelium in centre
colour	olive-black
reverse	black

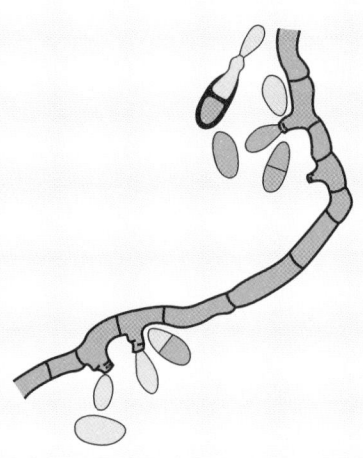

MICROSCOPIC APPEARANCE
at 30°C

predominant features	mostly yeast-like at first; later developing wide, septate hyphae with thick, brown walls; two-celled, pigmented, ellipsoidal conidia
conidiophore	wide, conspicous annellides formed on side projections of hyphae
conidia	colourless at first, becoming pale brown; ellipsoidal, 7-9.5 μm × 3.5-4.5 μm; one-celled, becoming two-celled with pigmented septa; often budding to form the yeast-like phase

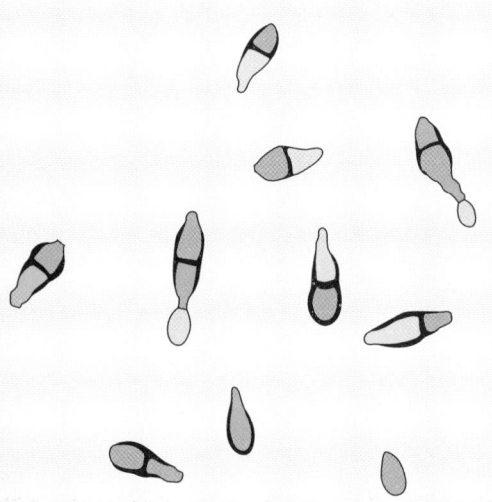

10 μm

DIFFERENTIAL DIAGNOSIS

colonial appearance   *Aureobasidium pullulans*, yeast-like phase of *Exophiala dermatitidis, E. jeanselmei*

microscopic appearance   *Exophiala* spp., but *P. werneckii* has broader annellated zones

SEXUAL STATE

None known.

CLINICAL IMPORTANCE

It is the cause of tinea nigra.

EXOPHIALA SPINIFERA

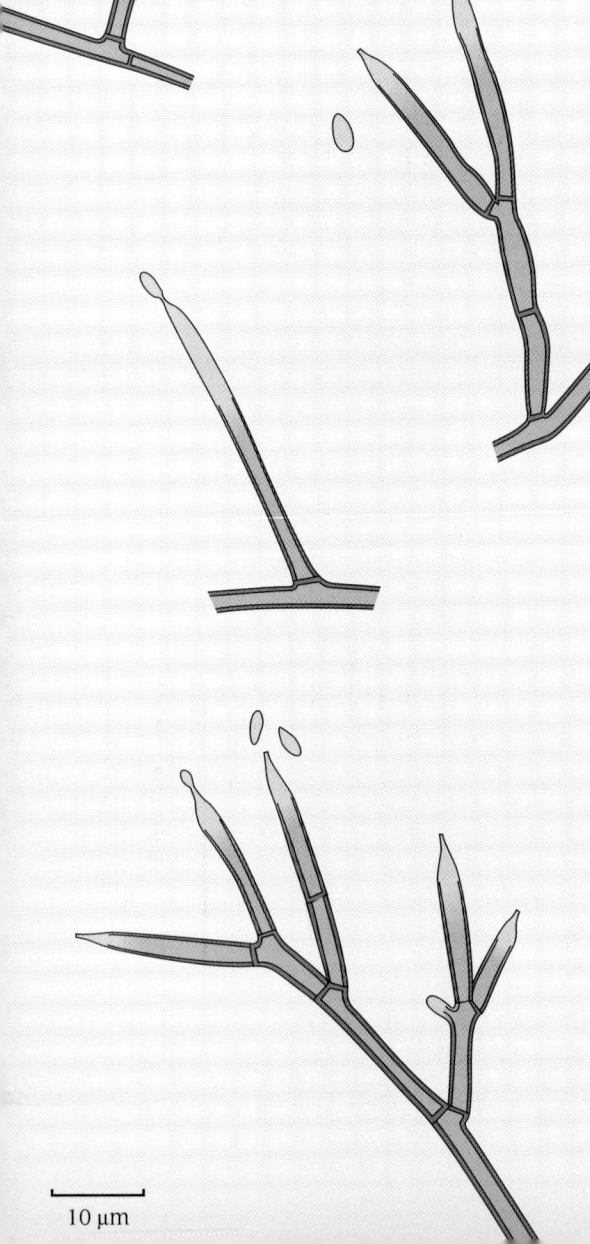

COLONIAL APPEARANCE
at 30°C on glucose peptone agar

diameter	10 mm in one week
topography	flat
texture	mucoid in centre; slightly floccose or velvety aerial mycelium developing at the edge of colonies
colour	olive to black
reverse	olive to black

MICROSCOPIC APPEARANCE
at 30°C

predominant features	abundant, budding, yeast-like cells; brown, septate hyphae with conidiophores of straight, brown annellides
conidiophore	cylindrical, brown, thick-walled branches, each with a spine-like annellated apical region
conidia	pale brown, oval to cylindrical, 1.8-2.8 µm × 2-4 µm

DIFFERENTIAL DIAGNOSIS

colonial appearance	other *Exophiala* spp., *Wangiella* spp., *Phialophora* spp., *Aureobasidium* spp., *Cladosporium* spp., *Scedosporium prolificans*
microscopic appearance	other *Exophiala* spp. but *E. spinifera* is distinguished by its spine-like conidiophores

SEXUAL STATE

None known.

CLINICAL IMPORTANCE

It is a cause of subcutaneons and deep forms of phaeohyphomycosis.

EXOPHIALA DERMATITIDIS

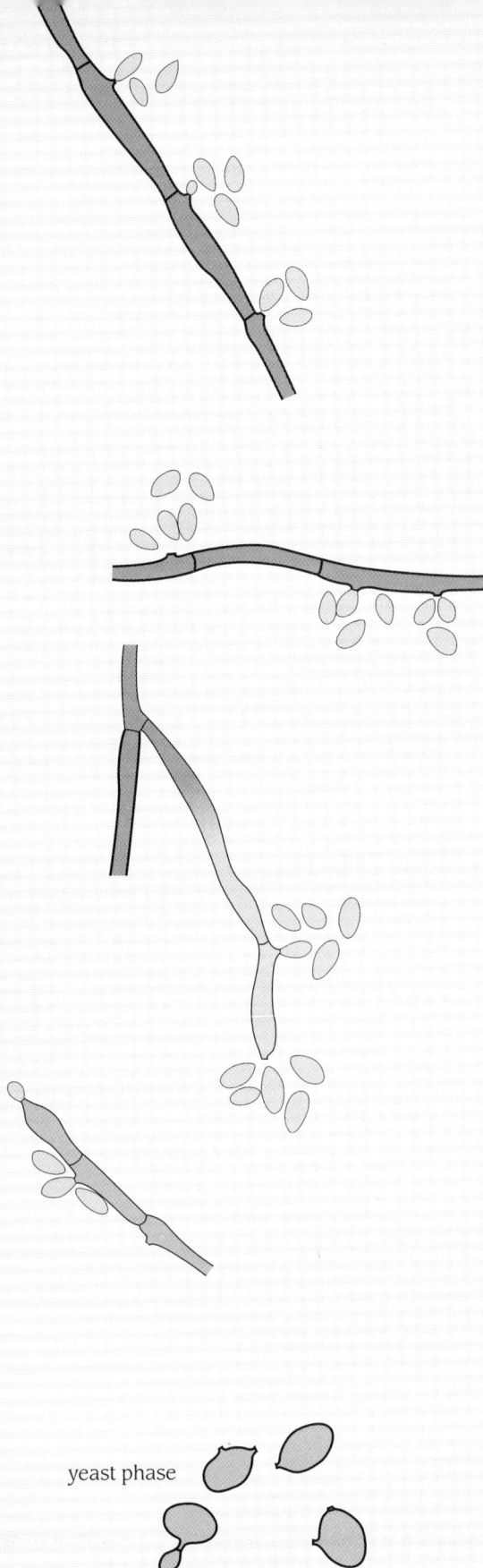

yeast phase

10 µm

COLONIAL APPEARANCE
at 30°C on glucose peptone agar

diameter	10 mm in one week
topography	flat
texture	mucoid, becoming velvety with the production of aerial mycelium
colour	olive-grey to black
reverse	olive-grey, often with diffusing, brown pigment

MICROSCOPIC APPEARANCE
at 30°C

predominant features	abundant, budding, yeast-like cells; few brown, septate hyphae with ellipsoidal conidia accumulating in groups
conidiophore	apical phialides without a collarette produced from hyphae or on short branches; annellated pores or pegs on the sides of hyphae
conidia	pale brown, ellipsoidal, 2.5-4 µm × 2-3 µm; accumulate in slimy clusters

DIFFERENTIAL DIAGNOSIS

colonial appearance other *Exophiala* spp., *Phialophora* spp., *Fonsecaea* spp., *Cladosporium* spp., *Scedosporium prolificans*

microscopic appearance other *Exophiala* spp.; *E. jeanselmei* except that *E. dermatitidis* grows well at 37°C and does not utilise nitrate

SEXUAL STATE

None known.

CLINICAL IMPORTANCE

It is a cause of subcutaneous and deep forms of phaeohyphomycosis. It may be found as a coloniser in the lungs of patients with cystic fibrosis.

EXOPHIALA JEANSELMEI

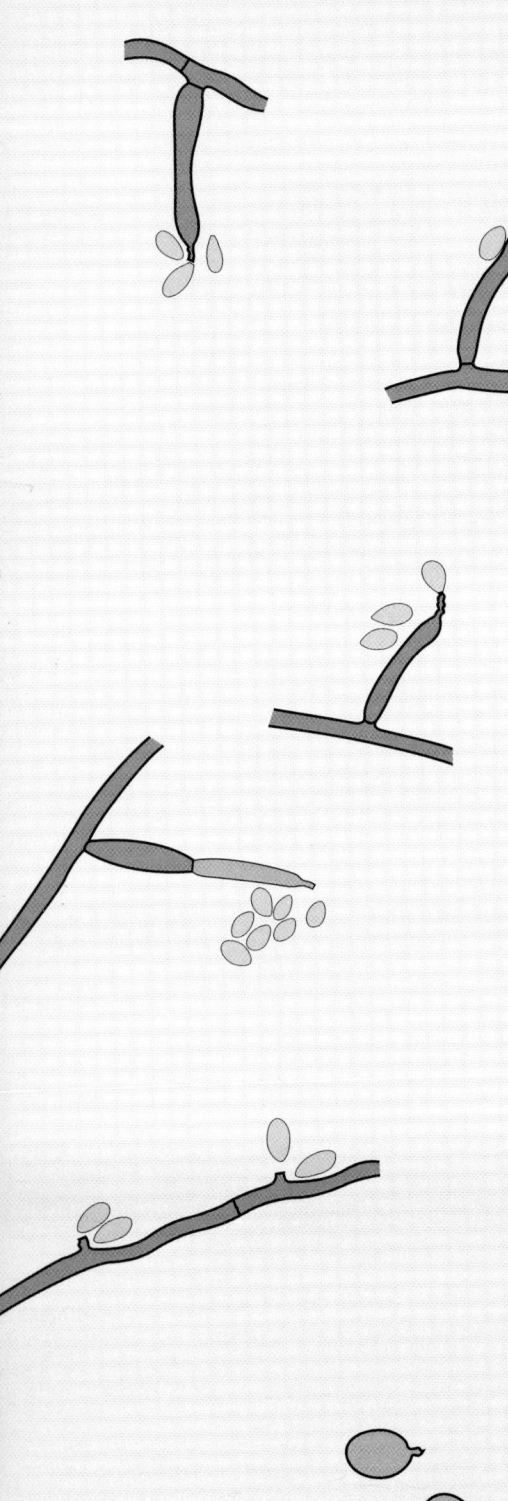

COLONIAL APPEARANCE
at 30°C on glucose peptone agar

diameter	10 mm in one week
topography	flat to domed
texture	mucoid, becoming overgrown with aerial mycelium
colour	dark olive-green to black
reverse	black

MICROSCOPIC APPEARANCE
at 30°C

predominant features	brown, septate hyphae; small, single-celled conidia accumulate in groups; mucoid colonies show many brown yeast cells with annellidic projections
conidiophore	brown, cylindrical to flask-shaped annellides with narrow apical area; borne on the sides of hyphae; annellidic sites also occur as peg-like outgrowths on long hyphae and on yeast cells
conidia	ellipsoidal, 2.6-5.9 μm × 1.2-2.5 μm, single-celled, colourless to pale brown

VARIANT FORMS

var. *lecanii-corni*   the annellidic sites are all peg-like and on the sides of undifferentiated hyphae

var. *heteromorpha*   the annellidic sites are all peg-like but on chains of round cells

DIFFERENTIAL DIAGNOSIS

colonial appearance   other *Exophiala* spp., *Wangiella* spp., *Phialophora* spp., *Fonsecaea* spp., *Cladosporium* spp., *Scedosporium prolificans*

microscopic appearance   other *Exophiala* spp.; *E. dermatitidis*, except that *E. jeanselmei* gives a positive nitrate test and is unable to grow at 40°C

SEXUAL STATE

None known.

CLINICAL IMPORTANCE

It is a cause of black-grain mycetoma in non-compromised individuals and can cause various forms of phaeohyphomycosis in immunocompromised patients.

9 MUCORACEOUS MOULDS AND THE[IR]

INTRODUCTION

With one exception, the moulds described in this chapter belong to a distinct fungal Class, the Zygomycetes. This is defined by the large, thick-walled zygospore that is formed by sexual fusion of two compatible mycelia. Another feature shared by these moulds is the broad, aseptate, vegetative hyphae that are unlike those of other fungi. In practice the zygospores are rarely seen, the asexual sporing structures serving to distinguish the genera and individual species.

Most of the species described here belong to the Order Mucorales. Members of this Order produce asexual spores in sacs known as sporangia, each of which is held above the substratum by a stalk or sporangiophore (see illustration opposite). In some species the stalk arises from a branched root-like system of rhizoids. The sporangium releases its contents by rupture of the outer wall, and this results in a ring-scar or frill where the wall separates from the stalk. The base of the sporangium is delineated by a dome-shaped septum, termed the columella. The size and shape of the columella of spent sporangia are often useful characteristics for identification. In contrast, in the genus *Cunninghamella*, the sporangiophore consists of numerous single-spored sporangioles arranged around the swollen end of the stalk.

These fungi are important saprophytes on decaying plant matter. Some are common as airborne sporangiospores and a few thermotolerant species cause the infection mucormycosis (zygomycosis) in humans and animals. Thus they may occur in the laboratory as pathogens or contaminants. An important character in eliminating the large number of non-pathogenic species is their inability to grow at blood temperature.

These moulds grow fast in culture, usually filling a standard petri dish in a few days. In many cases they produce a luxuriant aerial mycelium which occupies the entire air space of the plate. The sporangial structures are very delicate and easily disrupted on handling and therefore a combination of adhesive tape mounts and stereomicroscopic examination of the intact colonies is advisable before deciding on an identification.

Several species fail to produce sporing structures on normal media and require special substrates, such as Czapek-Dox agar, or pieces of agar/plant material floating in sterile tap water to induce sporulation. Even without this treatment, they can at least be assigned to the group by the broad, aseptate, vegetative hyphae.

The other great division of the Zygomycetes, the Order Entomophthorales, is represented here by the genera *Basidiobolus* and *Conidiobolus*. These fungi are associated with insects or amphibia and are not normally thought of as contaminants.

In some tropical regions they cause subcutaneous entomophthoramycosis. They reproduce asexually by large single-celled conidia which are forcibly ejected into the air by an explosive process. Growth is slow with little aerial mycelium and results in wrinkled, membranous colonies. The discharged conidia can be found adhering to and germinating on the lid of the plate.

Finally, *Pythium insidiosum* is a member of the Class Oomycetes or 'water moulds', and causes subcutaneous pythiosis in wet tropical environments. Colonies on agar are slow-growing, submerged and without sporing structures. Production of the characteristic flagellated zoospores requires culture on plant material floating in sterile tap water.

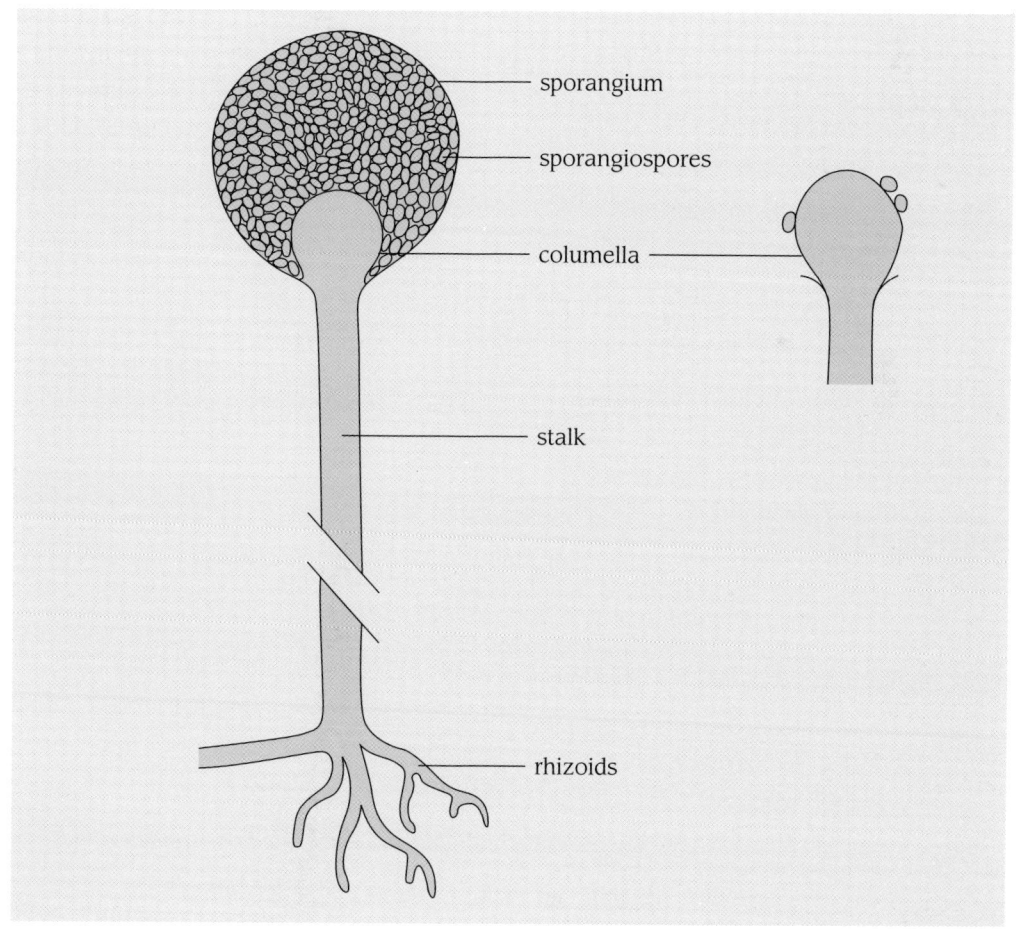

Key to species sporulating on glucose peptone agar

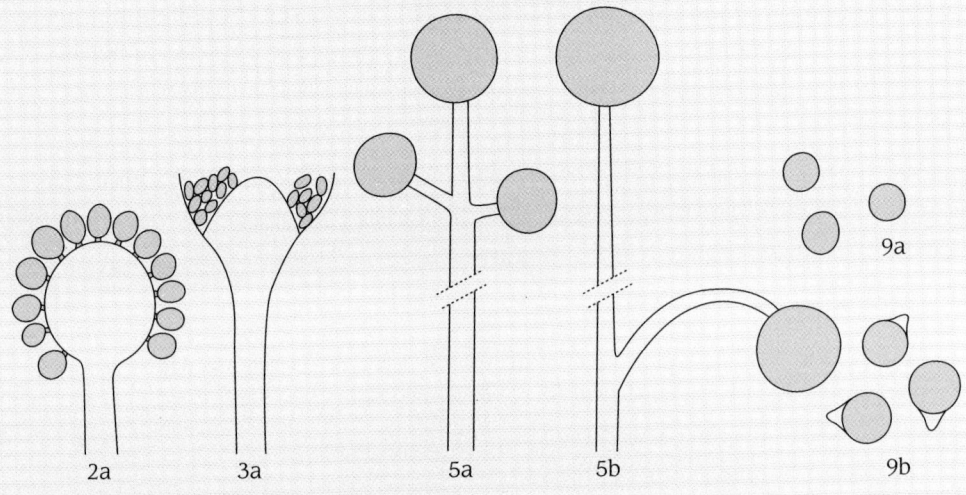

1a	Colony floccose; fast growth at 37°C	2	
1b	Colony floccose; no growth at 37°C	7	
1c	Colony slow growing, membranous and waxy	9	
2a	Spores produced outside a vesicle	*Cunninghamella bertholletiae*	
2b	Spores produced inside sporangia	3	
3a	Stalks with funnel-shaped top	*Absidia corymbifera*	
3b	Stalks not expanded at top	4	
4a	Stalks with a few side branches	5	
4b	Stalks unbranched	6	
5a	Branches crowded near top of main stalk	*Rhizomucor pusillus*	
5b	Branches more widely spaced; often curved	*Mucor circinelloides*	
6a	Stalks about 500 µm long; spores 4-6 µm	*Rhizopus microsporus*	
6b	Stalks over 1000 µm long; spores 6-8 µm	*Rhizopus arrhizus*	
7a	Sporangia black; rhizoids prominent at base of stalk	*Rhizopus stolonifer*	
7b	Sporangia pale or brownish; rhizoids absent	8	
8a	Colony pale yellow	*Mucor hiemalis*	
8b	Colony pale brown; some stalks with chlamydospores	*Mucor racemosus*	
9a	Spores without conical papillae	*Basidiobolus ranarum*	
9b	Spores with conical papillae	*Conidiobolus coronatus*	

Key to species not sporing on glucose peptone agar

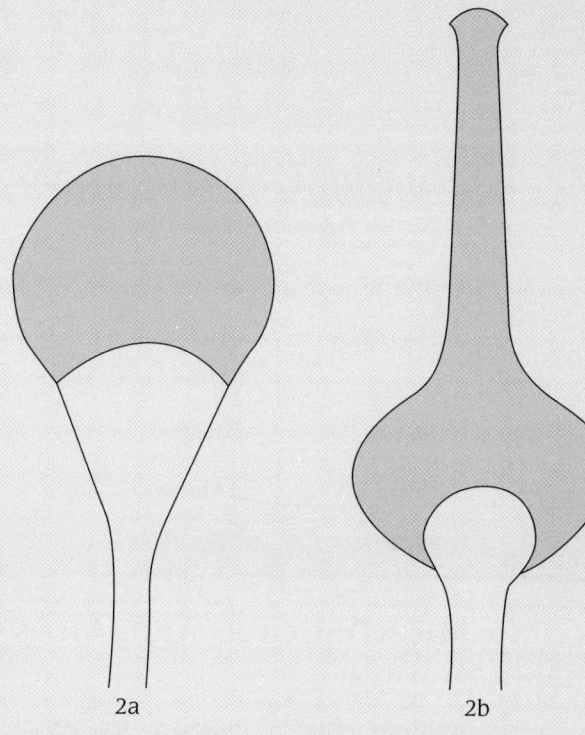

1a Motile zoospores produced in liquid culture   *Pythium* spp.
1b Mucoraceous sporangia produced on special media   2

2a Sporangia on funnel-shaped bases   *Apophysomyces elegans*
2b Sporangia with apical tubular extension   *Saksenaea vasiformis*
2c Sporangia rupturing without leaving a columella   *Mortierella wolfii*

CUNNINGHAMELLA BERTHOLLETIAE

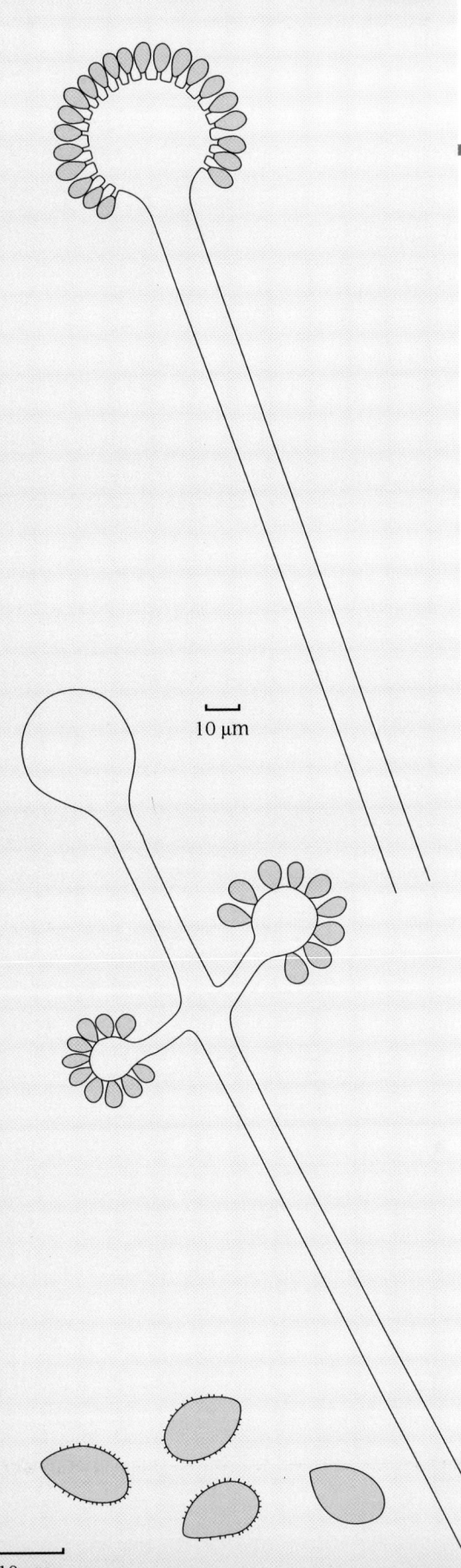

COLONIAL APPEARANCE
at 30°C on glucose peptone agar

diameter	90 mm in one week
topography	abundant aerial growth to lid of petri dish
texture	floccose
colour	white to pale grey
reverse	colourless

MICROSCOPIC APPEARANCE
at 30°C

predominant features	round to oval vesicles with spores on short stalks covering the entire surface
sporangiophore	round to slightly oval terminal vesicle formed at the end of a long, straight stalk; smaller vesicles sometimes occur at the ends of a whorl of branches just below the apical region
sporangiospores	oval, 7-11 µm, smooth or finely roughened; produced on short pegs over entire surface of vesicle

DIFFERENTIAL DIAGNOSIS

colonial appearance   other mucoraceous fungi

microscopic appearance   *Cunninghamella elegans*, but this does not grow at 45°C

SEXUAL STATE

Seldom found; spherical, brownish zygospores 25-55 µm in diameter with short projections.

CLINICAL IMPORTANCE

It is a rare cause of human disease.

ABSIDIA CORYMBIFERA

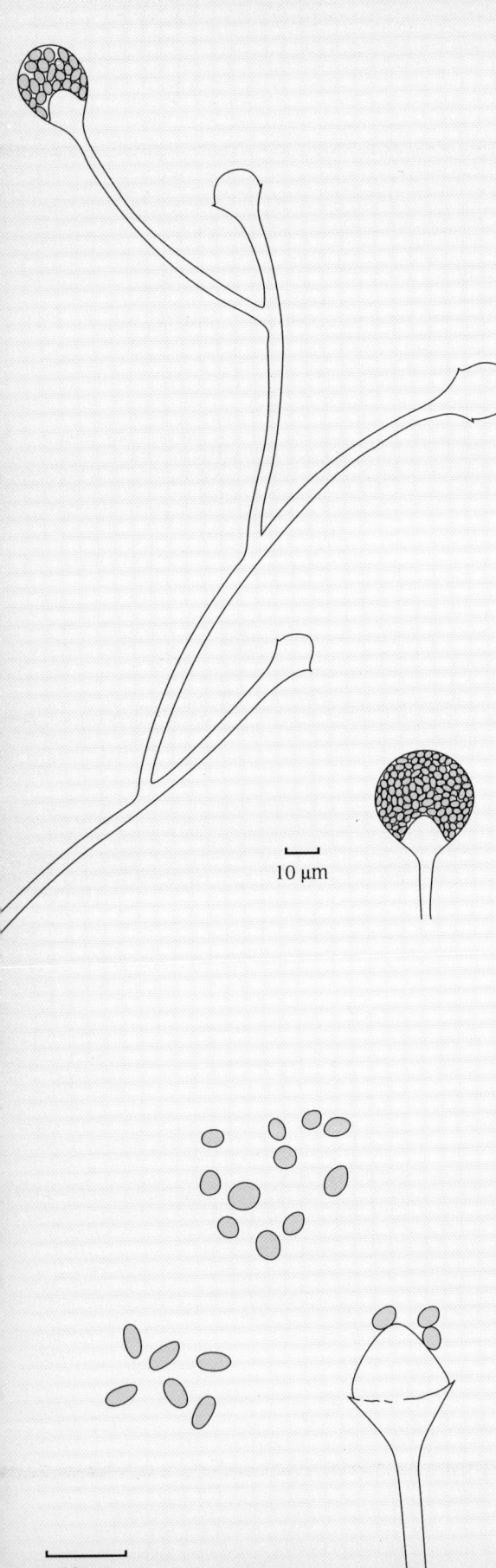

COLONIAL APPEARANCE
at 30°C on glucose peptone agar

diameter	90 mm in one week
topography	abundant aerial growth to lid of petri dish
texture	floccose
colour	white to pale grey
reverse	colourless

MICROSCOPIC APPEARANCE
at 30°C

predominant features	branching sporangiophores; numerous, small, pale sporangia with funnel-shaped bases
sporangiophore	mostly branched, sometimes in whorls; sporangium delineated by an almost conical columella
sporangiospores	ellipsoidal, 2-3.5 μm × 3-5 μm

DIFFERENTIAL DIAGNOSIS

colonial appearance	other mucoraceous fungi
microscopic appearance	*Mucor* spp., *Rhizopus* spp. and *Rhizomucor* spp., but *Absidia corymbifera* can be differentiated by the shape of the sporangium and the lack of rhizoids at the base of the sporangiophore

SEXUAL STATE

Seldom found; short, ellipsoidal, thick-walled, reddish-brown zygospores with marked ridges.

CLINICAL IMPORTANCE

It is a frequent cause of mucormycosis in immunocompromised patients. It is the most common aetiological agent of bovine mycotic abortion.

RHIZOMUCOR PUSILLUS

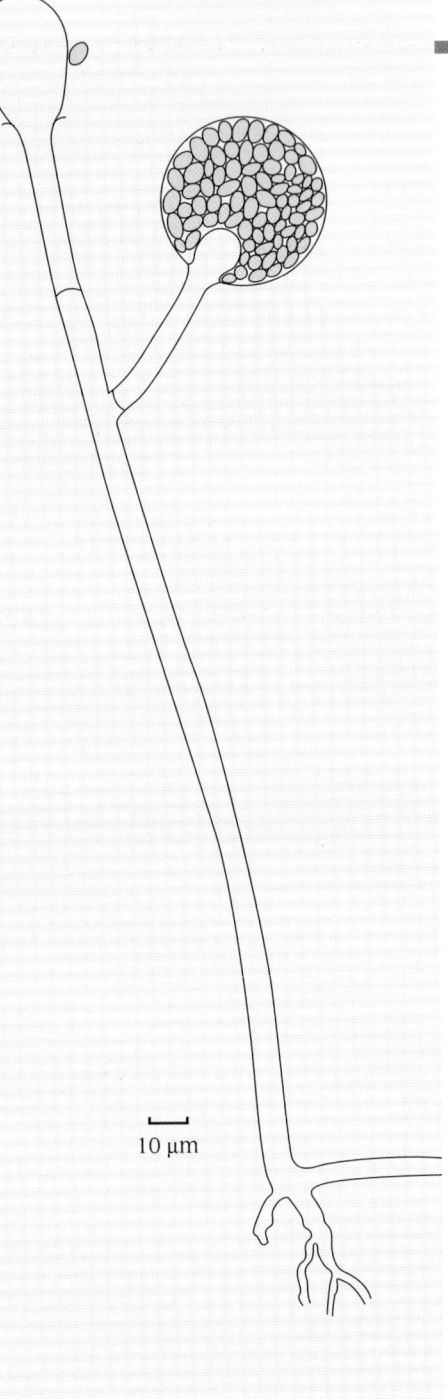

COLONIAL APPEARANCE
at 30°C on glucose peptone agar

diameter	90 mm in one week
topography	flat with low aerial growth
texture	floccose
colour	brownish grey
reverse	colourless

MICROSCOPIC APPEARANCE
at 30°C

predominant features	brown-pigmented sporangia in clusters on short branches of the main stalk; short, thin-walled rhizoids may occur at the base
sporangiophore	brownish stalk with branching near tip; small, brown, round sporangia up to 100 µm in diameter; sporangia produced at tips of branches with a septum in the branch
sporangiospores	small, 3-4 µm, colourless, smooth, round or almost round

DIFFERENTIAL DIAGNOSIS

colonial appearance   other mucoraceous fungi

microscopic appearance   *Mucor* spp., but *Rhizomucor pusillus* will tolerate temperatures up to 55°C

SEXUAL STATE

Seldom found; blackish brown, round zygospores, 70 μm in diameter, with roughened surface.

CLINICAL IMPORTANCE

It is an uncommon cause of human mucormycosis. It is an aetiological agent of bovine mycotic abortion.

MUCOR CIRCINELLOIDES

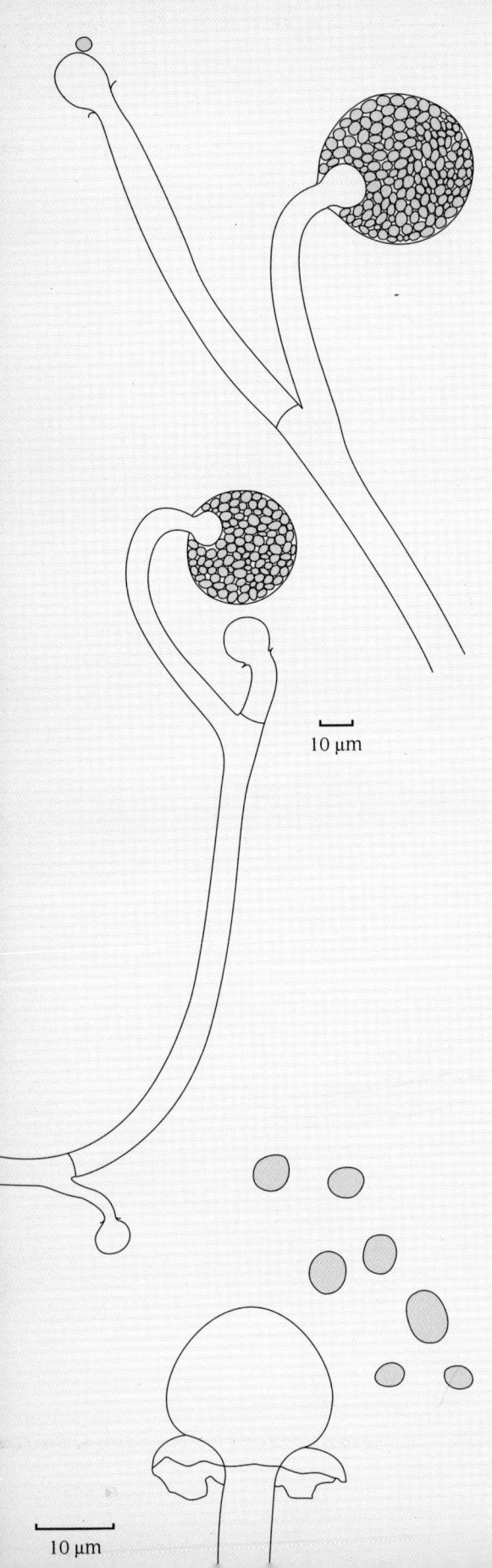

COLONIAL APPEARANCE
at 30°C on glucose peptone agar

diameter	90 mm in one week
topography	abundant aerial growth to lid of petri dish
texture	floccose
colour	pale yellowish brown
reverse	colourless

MICROSCOPIC APPEARANCE
at 30°C

predominant features	pale brown sporangia; occasional curved stalks
sporangiophore	a mixture of long, unbranched stalks and sympodially branching, shorter stalks which sometimes curve downwards; sporangia 20-80 µm in diameter with slightly encrusted walls
sporangiospores	smooth, ellipsoidal, up to 7 µm × 5 µm

DIFFERENTIAL DIAGNOSIS

colonial appearance   other mucoraceous fungi

microscopic appearance   other *Mucor* spp., except that *M. circinelloides* has curved, branching sporangiophores

Absidia, *Rhizopus* and *Rhizomucor* spp. except for the absence of rhizoids and poor growth at 37°C in *M. circinelloides*

SEXUAL STATE

Seldom found; reddish-brown to dark brown zygospores with spines, sub-globose, up to 100 µm in diameter.

CLINICAL IMPORTANCE

It is an occasional human pathogen.

RHIZOPUS MICROSPORUS

COLONIAL APPEARANCE
at 30°C on glucose peptone agar

diameter	90 mm in one week
topography	abundant aerial growth to lid of petri dish
texture	floccose
colour	grey to dark grey-brown
reverse	colourless

MICROSCOPIC APPEARANCE
at 30°C

predominant features	small, grey-black sporangia formed on short stalks arising from dark brown, branched rhizoids
sporangiophore	short, produced singly or in groups up to four; arising from a knot of rhizoids; small sporangia, up to 100 µm in diameter, grey-black with semi-spherical columellae
sporangiospores	small, up to 6 µm, round to slightly angular, striated

VARIANT FORMS

var. *rhizopodiformis*   has elongated columellae; spores less obviously striated

var. *oligosporus*   has larger spores, 7-10 μm, without striations

DIFFERENTIAL DIAGNOSIS

colonial appearance   other mucoraceous fungi

microscopic appearance   *Rhizopus arrhizus*, but *R. microsporus* is distinguished by its shorter stalks and smaller sporangia and spores

SEXUAL STATE

Seldom found; round zygospores, up to 100 μm in diameter, reddish brown with surface projections.

CLINICAL IMPORTANCE

It causes human mucormycosis and bovine mycotic abortion.

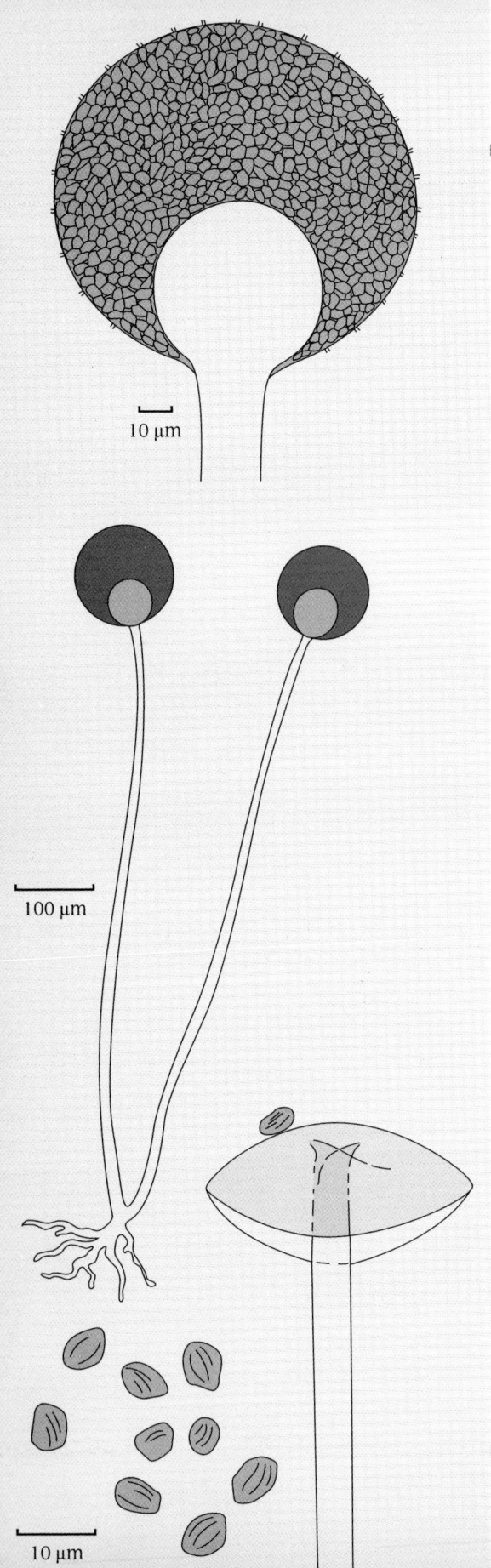

RHIZOPUS ARRHIZUS

COLONIAL APPEARANCE
at 30°C on glucose peptone agar

diameter	90 mm in one week
topography	abundant aerial growth to lid of petri dish
texture	floccose
colour	pale-grey to greyish brown, with black sporangia near edge of plate
reverse	colourless

MICROSCOPIC APPEARANCE
at 30°C

predominant features	wide, non-septate, colourless hyphae; brownish rhizoids; brownish black, spherical sporangia with large columellae
sporangiophore	stalks single or in groups arising from rhizoids; brownish black, spherical sporangia, 50-250 µm in diameter; large columellae collapse giving appearance of mushrooms
sporangiospores	greyish green, variable in shape, 6-8 µm long; angular with longitudinal striations

DIFFERENTIAL DIAGNOSIS

colonial appearance   other mucoraceous fungi

microscopic appearance   other *Rhizopus* spp.; *Mucor, Rhizomucor* and *Absidia* spp., all of which lack nodal rhizoids and collapsed columellae

SEXUAL STATE

Seldom found; red to brown, spherical zygospores with flat projections.

CLINICAL IMPORTANCE

It is the principal aetiological agent of human mucormycosis, causing rhinocerebral, lung, gastrointestinal, cutaneous or disseminated infection in predisposed individuals, such as diabetics and neutropenic patients. The different clinical forms are often associated with particular underlying conditions.

RHIZOPUS STOLONIFER

This environmental organism is very similar to *R. arrhizus*, but is larger in all respects (sporangia up to 350 µm diameter; sporangiospores 9-11 µm long). It has a maximum growth temperature of 32°C and has not been reported to cause human infection.

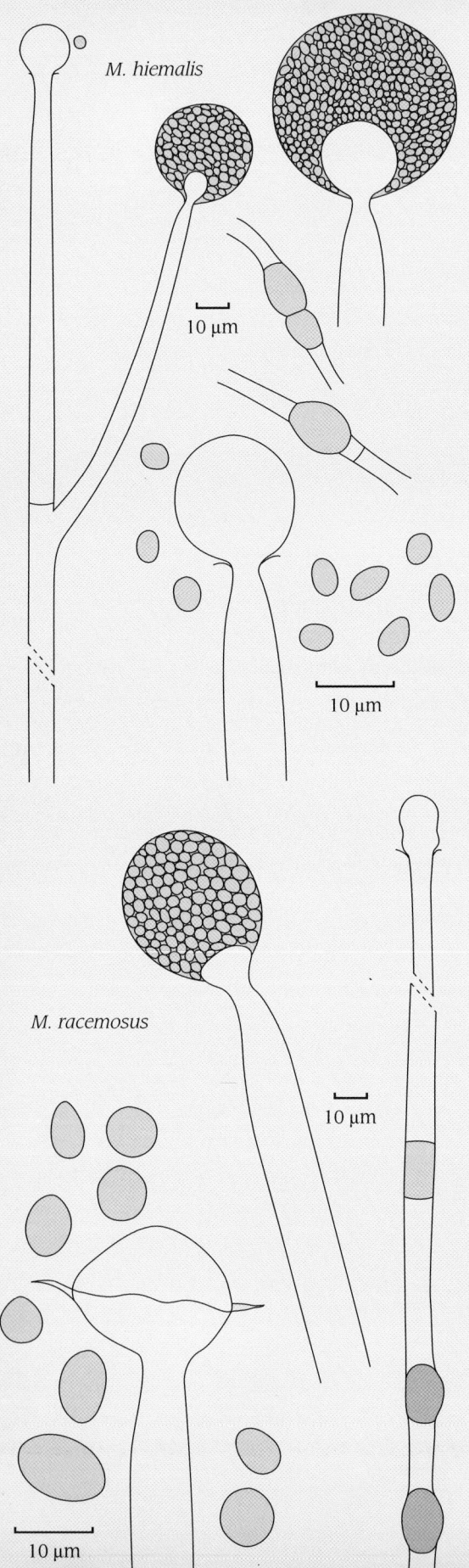

MUCOR HIEMALIS

COLONIAL APPEARANCE
at 30°C on glucose peptone agar

diameter	90 mm in one week
topography	abundant aerial growth to lid of petri dish
texture	floccose
colour	pale yellow
reverse	colourless

MICROSCOPIC APPEARANCE
at 30°C

predominant features	dark brown, spherical sporangia; chlamydospores in hyphae but not in sporangiophores
sporangiophore	unbranched at first, later sparingly branched; yellow sporangia becoming dark brown, 20-80 µm
sporangiospores	smooth, ellipsoidal, flattened at one side, variable in size up to 9 µm × 5.5 µm

DIFFERENTIAL DIAGNOSIS

colonial appearance	other mucoraceous fungi
microscopic appearance	other *Mucor* and *Absidia* spp., but *M. hiemalis* will not grow above 30°C

SEXUAL STATE

Seldom found; black-brown zygospores up to 100 μm in diameter with long spines.

CLINICAL IMPORTANCE

This common environmental organism is a very rare cause of human infection.

MUCOR RACEMOSUS

This common environmental organism is very similar to other *Mucor* spp. but can be distinguished by the presence of chlamydospores occurring in the sporangiophores. It has a maximum growth temperature of 32°C and has not been reported to cause human infection.

BASIDIOBOLUS RANARUM

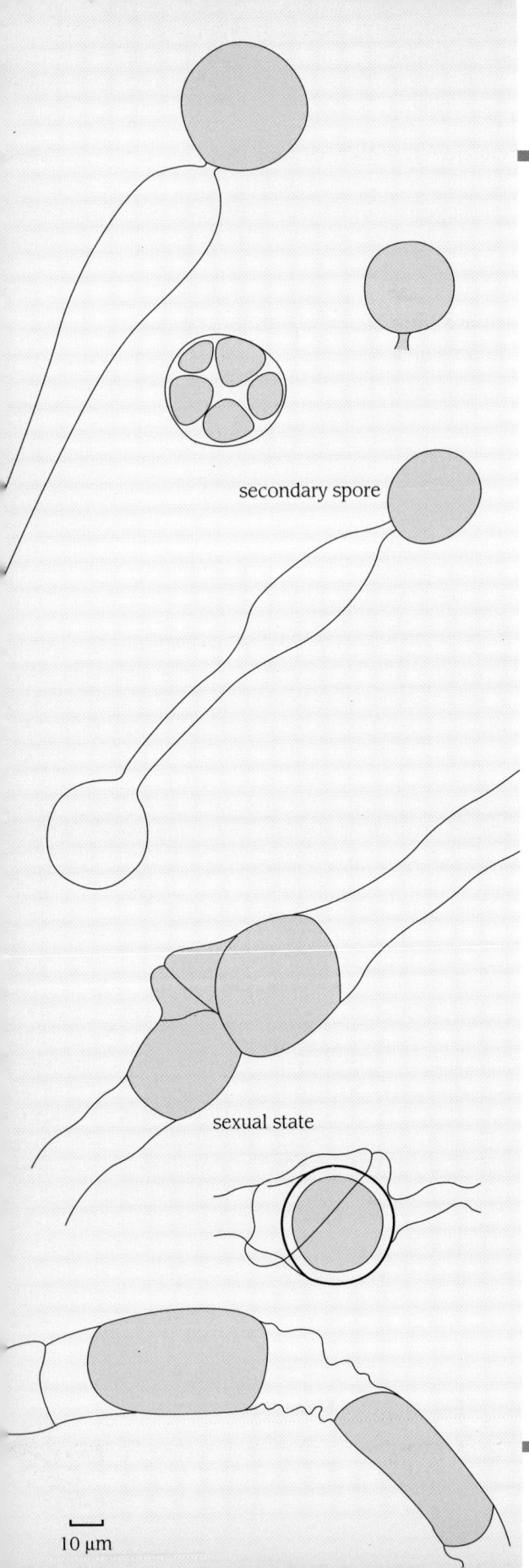

secondary spore

sexual state

10 μm

COLONIAL APPEARANCE
at 30°C on glucose peptone agar

diameter	20 mm in one week
topography	flat, wrinkled or radially folded
texture	waxy and glabrous with very short aerial mycelium
colour	yellow-grey to buff
reverse	buff

MICROSCOPIC APPEARANCE
at 30°C

predominant features	wide hyphae with few septa when young; older cultures contain large, round chlamydospores and colourless zygospores; ballistospores forcibly discharged on to lid of petri dish
sporophore	primary spores formed singly on swollen ends of unbranched hyphae
spores	round to pear-shaped, up to 40 μm in diameter; a secondary spore may develop from a swollen hypha emerging from the germinated spore, or the latter may develop a number of endospores

DIFFERENTIAL DIAGNOSIS

colonial appearance *Conidiobolus coronatus*

microscopic appearance *C. coronatus*, in which swelling of the sporophore below the primary spore and beaked zygospores are absent

SEXUAL STATE

Round, smooth, thick-walled zygospores, 20-50 µm in diameter, with prominently beaked cells attached.

CLINICAL IMPORTANCE

It is the cause of a chronic subcutaneous infection of the trunk and limbs which occurs in tropical regions in East or West Africa and South East Asia.

CONIDIOBOLUS CORONATUS

COLONIAL APPEARANCE
at 30°C on glucose peptone agar

diameter	80 mm in one week
topography	wrinkled or with radial folds
texture	glabrous or waxy at first; becoming powdery with age
colour	white; becoming buff with age
reverse	white

MICROSCOPIC APPEARANCE
at 30°C

predominant features	wide, septate hyphae; large, round spores forcibly discharged on to lid of petri dish
sporophore	primary spores formed singly on slightly tapering ends of unbranched hyphae
spores	round primary spores may germinate after discharge to produce either single or multiple secondary spores on short stalks; hair-like appendages may be formed on some spores; others may germinate by hyphal outgrowth

secondary spores

spore with hair-like outgrowths

geminating spore

10 μm

DIFFERENTIAL DIAGNOSIS

colonial appearance *Basidiobolus* spp.

microscopic appearance *Basidiobolus* spp., but *Conidiobolus coronatus* lacks the swelling of the sporophore just below the spore, and the multiple secondary spores

SEXUAL STATE

None known.

CLINICAL IMPORTANCE

It is the cause of a chronic rhinofacial infection which occurs in the tropical rain forests of Africa and Central and South America.

PYTHIUM INSIDIOSUM

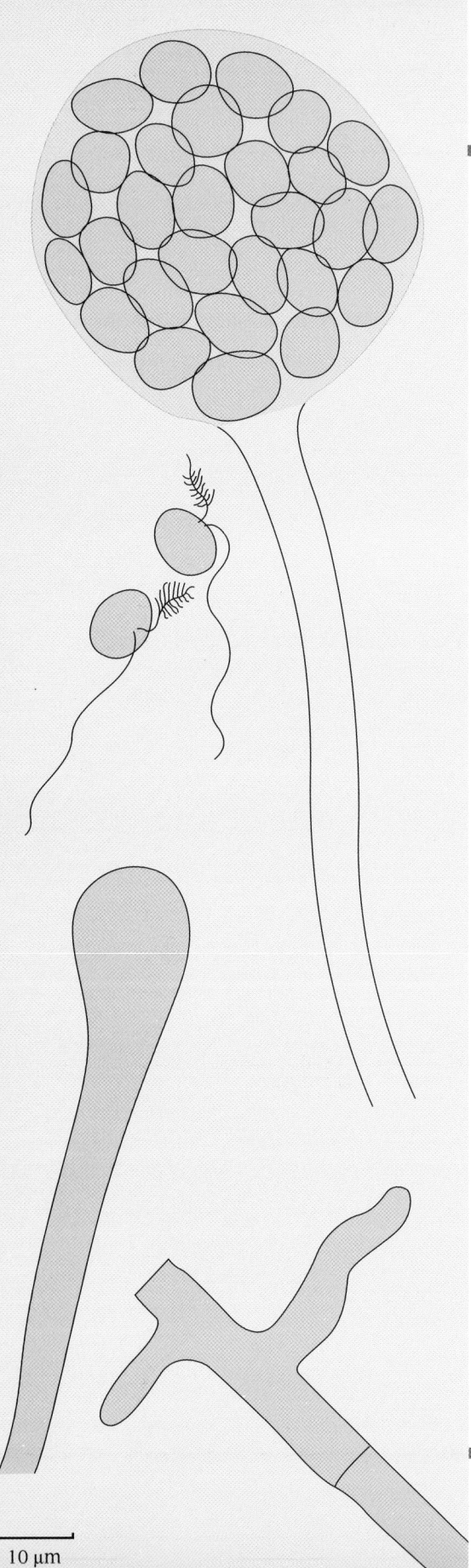

COLONIAL APPEARANCE
at 30°C on glucose peptone agar

diameter	90 mm in one week
topography	undulated, radiating
texture	submerged hyphae or very short aerial mycelium
colour	white to yellowish white
reverse	colourless

MICROSCOPIC APPEARANCE
at 30°C

predominant features	wide, irregularly septate hyphae of varying width with large, round swellings; hyphae break easily at the septa
sporangiophore	motile zoospores develop from a mass of cytoplasm extruded through the expanded tip of an undifferentiated hypha
zoospores	round to oval, 7 µm × 10 µm, with two lateral flagella

Note: sporulation requires culture on agar blocks floated in water.

DIFFERENTIAL DIAGNOSIS

colonial appearance   *Basidiobolus ranarum, Conidiobolus coronatus*

microscopic appearance   other non-sporing zygomycetes (on glucose peptone agar)

SEXUAL STATE

Subspherical, intercalary oogonia may be produced on some agar media.

CLINICAL IMPORTANCE

There have been a few reports of subcutaneous infection in humans. This organism causes pythiosis in animals, for example 'swamp cancer' in horses in tropical regions.

APOPHYSOMYCES ELEGANS

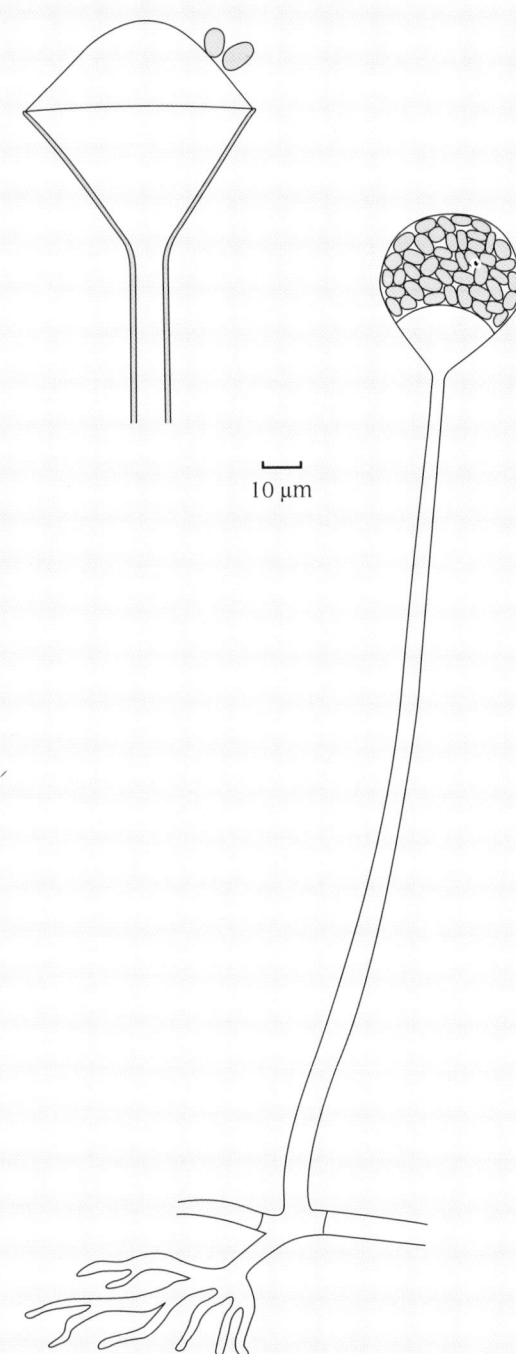

COLONIAL APPEARANCE
at 30°C on glucose peptone agar

diameter	90 mm in one week
topography	abundant aerial growth to lid of petri dish
texture	floccose
colour	white becoming cream to pale yellow
reverse	colourless

MICROSCOPIC APPEARANCE
at 30°C

predominant features	wide, non-septate, colourless hyphae; on some media it forms unbranched, solitary sporangiophores with thin-walled, colourless to greyish brown rhizoids
sporangiophore	solitary, straight or slightly curved, unbranched, colourless to greyish brown; pear-shaped sporangia produced terminally with prominent, bell- or funnel-shaped enlargement of sporangiophore (apophysis) just below the semi-spherical columella
sporangiospores	smooth, slightly brown, oval, 5-8 µm × 4-6 µm

DIFFERENTIAL DIAGNOSIS

colonial appearance   *Mucor* spp., *Rhizopus* spp., *Rhizomucor* spp., *Absidia* spp.

microscopic appearance   *Absidia* spp., but *Apophysomyces elegans* shows distinctive apophyses and unbranched, straight stalks

SEXUAL STATE

None known.

CLINICAL IMPORTANCE

It is an uncommon cause of mucormycosis.

SAKSENAEA VASIFORMIS

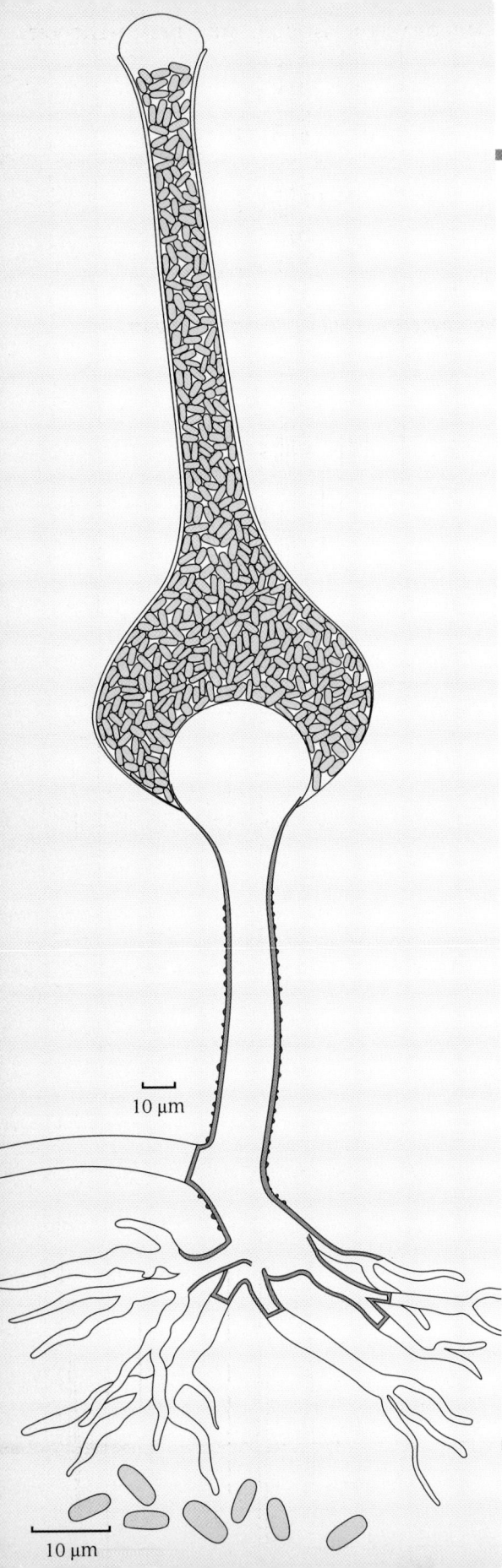

COLONIAL APPEARANCE
at 30°C on glucose peptone agar

diameter	90 mm in one week
topography	abundant aerial growth to lid of petri dish
texture	floccose
colour	white to pale grey
reverse	colourless

MICROSCOPIC APPEARANCE
at 30°C

predominant features	distinctive, long, flask-shaped sporangia; brown rhizoids
sporangiophore	arises singly from dark brown rhizoids; short, thick-walled, with a very long flask-shaped sporangium
sporangiospores	small, up to 4 µm × 2 µm, smooth, cylindrical; released through opening at tip of sporangium

Note: sporulation requires culture on special media, e.g. Czapek-Dox agar.

DIFFERENTIAL DIAGNOSIS

colonial appearance   other mucoraceous fungi

microscopic appearance other non-sporing zygomycetes (on glucose peptone agar)

SEXUAL STATE

None known.

CLINICAL IMPORTANCE

It is an uncommon cause of mucormycosis.

MORTIERELLA WOLFII

COLONIAL APPEARANCE
at 30°C on glucose peptone agar

diameter	90 mm in one week
topography	flat, with irregular edge
texture	thinly floccose
colour	grey to greyish yellow
reverse	colourless

MICROSCOPIC APPEARANCE
at 30°C

predominant features	wide, colourless, non-septate hyphae; chlamydospores with blunt-ended wall outgrowths may occur
sporangiophore	arises from rhizoids; tapers towards the tip and forms a whorl of branches in the apical region; discharge of sporangia leaves a frill without a columella
sporangiospores	ellipsoidal to kidney-shaped, up to 12 µm × 6 µm, smooth-walled, colourless

Note: sporulation requires culture on special media, e.g. Czapek-Dox agar.

DIFFERENTIAL DIAGNOSIS

colonial appearance   other mucoraceous fungi

microscopic appearance other non-sporing zygomycetes (on glucose peptone agar)

SEXUAL STATE

None known.

CLINICAL IMPORTANCE

It is a cause of bovine mycotic abortion but has not been reported to cause human disease.

10 MISCELLANEOUS MOULDS

INTRODUCTION

The moulds described in this chapter include a number that produce macroscopic fruiting bodies in culture. These bodies may be pycnidia, which contain asexual conidia (pycnidiospores); ascocarps, which contain ascospores; or basidiocarps, which produce basidiospores. In addition, Madurella, a genus of non-sporing moulds which cause black-grain mycetoma, has been included here, as it cannot be accommodated in earlier chapters based on types of conidiogenesis.

Of the large number of fungi capable of producing fruiting bodies in culture, only those with a well-established pathogenic potential have been included. It should also be noted that a number of moulds that develop fruiting bodies have already been described by conidial type, as the conidia are usually the dominant spore state present. Descriptions of the ascocarps of *Aspergillus glaucus, A. nidulans* and *Scedosporium apiospermum*, and the pycnidia of *Scytalidium dimidiatum* will therefore be found in earlier chapters.

Although the fruiting bodies of some fungi are readily formed on normal media in one or two weeks, others require prolonged incubation for one to two months on special media. Individual species differ in their requirements for light, temperature and type of substrate and it is possible to obtain by experiment conditions that stimulate the development and maturation of fruiting bodies. In clinical practice this effort is only warranted when there is good clinical and/or histopathological evidence of fungal infection.

In studying the fruiting bodies, a plate microscope with top and transmitted illumination should be used to establish their position relative to the agar surface, to detect the presence or absence of an ostiole through which the spores are liberated and to investigate the nature of any specialised hyphae growing from the structure. The fruiting bodies should then be carefully dissected from the colonies and mounted under a cover slip for measurement of their diameter and crushing to reveal their contents. Some fruiting bodies will be so hard that the cover slip will break if crushing is attempted. If so, the bodies should be crushed between two microscope slides before mounting the debris in the usual way. The details of the minute structures lining the wall of a pycnidium are best seen under oil immersion using an aqueous mounting fluid such as 1% eosin, rather than lactofuchsin or lactophenol.

Key to miscellaneous moulds

1a	Fruiting bodies produced within two weeks	2
1b	Fruiting bodies produced after two weeks	5
1c	Fruiting bodies never produced	9
2a	Fruiting bodies with pale walls	3
2b	Fruiting bodies with dark walls	4
3a	Colony white to buff-coloured	*Aphanoascus fulvescens*
3b	Colony dull purple	*Monascus ruber*
4a	Fruiting bodies are thick-walled ascocarps, covered in long, brown hyphae	*Chaetomium globosum*
4b	Fruiting bodies are pycnidia, with thin, smooth walls	*Phoma herbarum*
5a	Fruiting bodies consist of a loose network of dark hyphae	*Myxotrichum deflexum*
5b	Fruiting bodies with gills	*Schizophyllum commune*
5c	Fruiting bodies black, with a definite wall	6
6a	Fruiting bodies are ascocarps; contain asci	7
6b	Fruiting bodies are pycnidia	8
7a	Ascospores large, mostly five-celled	*Leptosphaeria senegalensis*
7b	Ascospores small, two-celled; ascocarp wall of interlocking plates	*Neotestudina rosatii*
7c	Ascospores long, flexuous, with tapered ends	*Piedraia hortae*
8a	Conidia large, two-celled when mature	*Lasiodiplodia theobromae*
8b	Conidia small, one-celled; pycnidia with spines	*Pyrenochaeta romeroi*
9a	Colony flat or folded, velvety, with a diffusing, brown pigment	*Madurella mycetomatis*
9b	Colony domed, densely floccose, without diffusing pigment	*Madurella grisea*

APHANOASCUS FULVESCENS

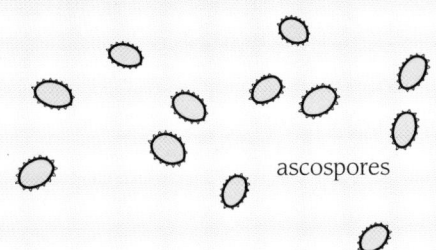
ascospores

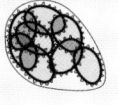

asci
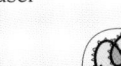

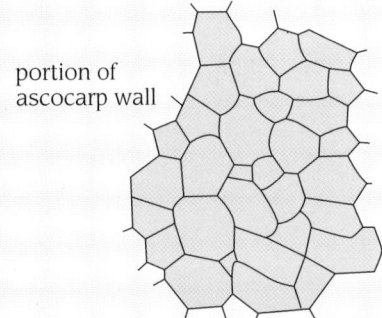
portion of ascocarp wall

10 μm

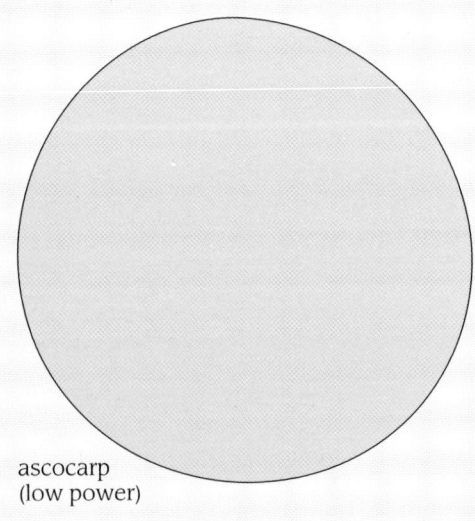
ascocarp (low power)

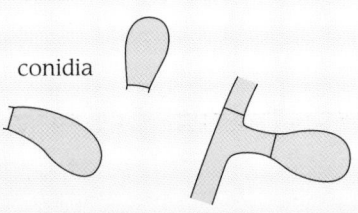
conidia

10 μm

COLONIAL APPEARANCE
at 30°C on glucose peptone agar

diameter	25 mm in one week
topography	flat
texture	powdery to felt-like
colour	white to cream
reverse	colourless

MICROSCOPIC APPEARANCE
at 30°C

predominant features	abundant, club-shaped conidia; older colonies develop round fruiting bodies near the centre
fruiting bodies	large ascocarps, 290-500 μm in diameter, colourless to light brown, with smooth, thick walls without an opening; on rupturing release oval to ellipsoidal asci, each containing eight ascospores
spores	ascospores light brown, lens- or disc-shaped, up to 5 μm × 3.5 μm, with markedly rough walls; conidia large, club-shaped, roughened, with a truncated base (see *Chrysosporium keratinophilum*)

DIFFERENTIAL DIAGNOSIS

colonial appearance   *Trichophyton* spp., *Chrysosporium* spp.

microscopic appearance   ascocarps of the Aspergillus group, but their conidial states differ from *Aphanoascus fulvescens*

ASEXUAL STATE

A. fulvescens is the sexual state of a *Chrysosporium* sp., very similar to *C. keratinophilum*.

CLINICAL IMPORTANCE

This keratinophilic soil organism is a rare cause of skin infections.

MONASCUS RUBER

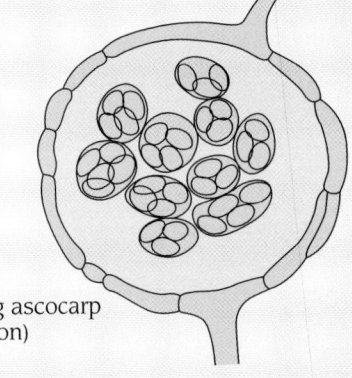

mature ascocarp (partial cross section)

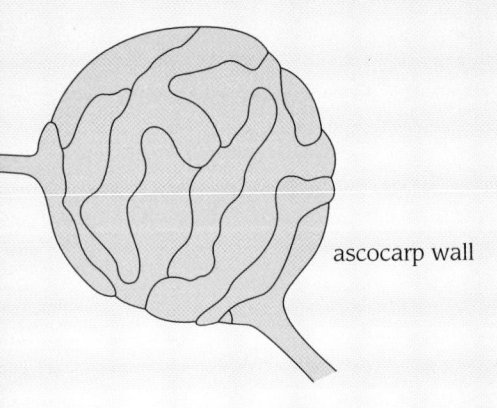

young ascocarp (section)

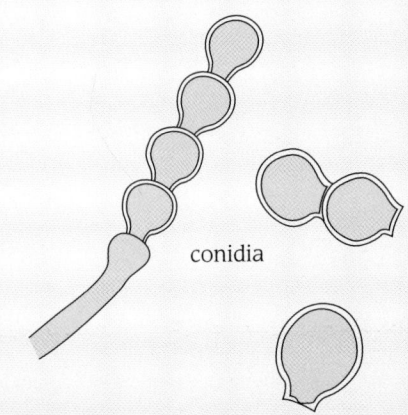

ascocarp wall

conidia

10 µm

COLONIAL APPEARANCE
at 30°C on glucose peptone agar

diameter	40 mm in one week
topography	flat, spreading
texture	thinly floccose
colour	reddish-grey to purple, becoming greyish
reverse	dull reddish-purple

MICROSCOPIC APPEARANCE
at 30°C

predominant features	numerous, round, thin-walled ascocarps; chains of conidia usually also present
fruiting bodies	brown, round, thin-walled ascocarps produced on short, stalk-like hyphae; mature ascocarps are packed with loose ascospores, but the asci may be seen clearly in the young ascocarps
spores	ascospores oval, 5.5-6 µm × 3.5-4 µm, smooth, colourless; chains of round, colourless conidia, 9-10.5 µm × 7-9 µm, with flattened bases, are also formed on undifferentiated hyphae

DIFFERENTIAL DIAGNOSIS

colonial appearance *Chaetomium* spp., *Ochroconis gallopava*, *Fusarium* spp.

microscopic appearance *Scopulariopsis brevicaulis* has similar chains of truncated conidia but no ascocarps

ASEXUAL STATE

Basipetospora rubra

CLINICAL IMPORTANCE

It is a rare cause of peritonitis and deep-seated infection.

CHAETOMIUM GLOBOSUM

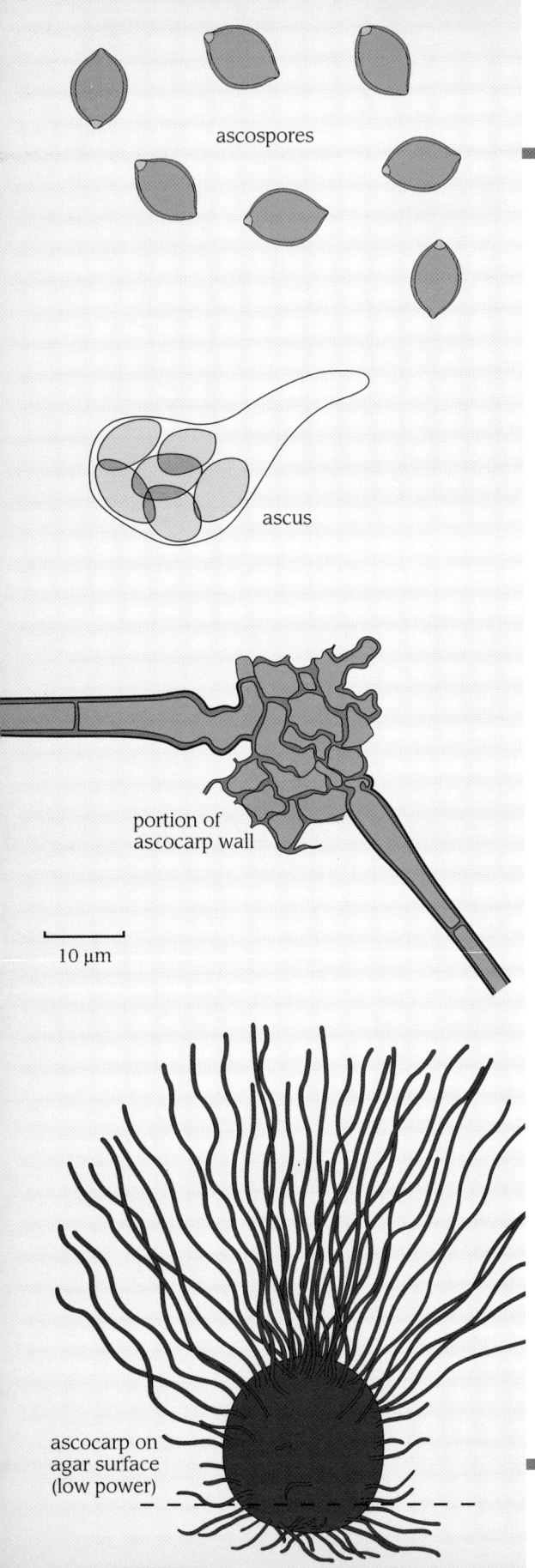

COLONIAL APPEARANCE
at 30°C on glucose peptone agar

diameter	30 mm in one week
topography	flat
texture	felt-like
colour	pale buff to olive-green
reverse	cream; sometimes with a diffusing, green, yellow or red pigment

MICROSCOPIC APPEARANCE
at 30°C

predominant features	large, brown ascocarps formed on surface of the agar; brown, lemon-shaped ascospores
fruiting bodies	oval ascocarps, up to 280 μm in diameter, with a terminal opening; dark brown ascocarp wall composed of a network of interwoven hyphae with unbranched, wavy, dark-coloured hyphae radiating from it; ascocarps contain club-shaped asci, each with eight ascospores
ascospores	pale brown, lemon-shaped, 9-12 μm × 6-8 μm, with an apical pore

DIFFERENTIAL DIAGNOSIS

colonial appearance the pycnidia of *Pyrenochaeta* spp. can resemble the ascocarps of *Chaetomium globosum*

microscopic appearance other *Chaetomium* spp.

CLINICAL IMPORTANCE

It is a common saprophyte on cellulosic materials, including paper. It is a rare cause of human disease.

PHOMA HERBARUM

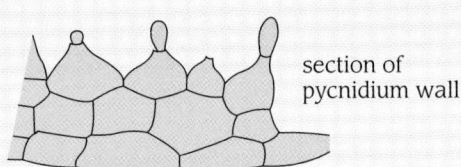
section of pycnidium wall

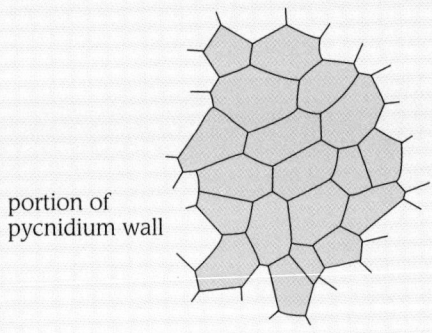
portion of pycnidium wall

10 μm

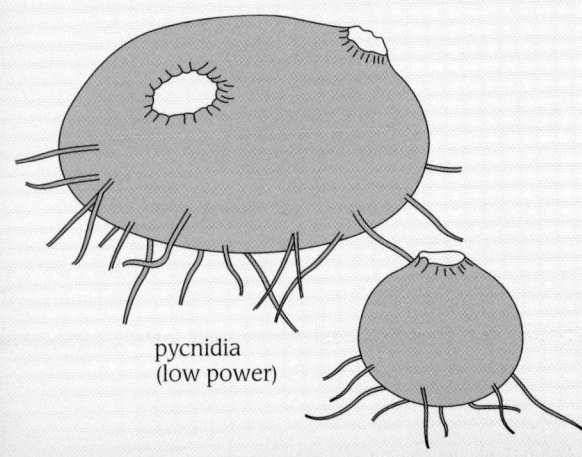
pycnidia (low power)

COLONIAL APPEARANCE
at 30°C on glucose peptone agar

diameter	25 mm in one week
topography	flat
texture	thinly floccose, over a wet basal layer
colour	reddish brown
reverse	pale brown

MICROSCOPIC APPEARANCE
at 30°C

predominant features	light brown pycnidia; colourless to pale pink conidial masses
fruiting bodies	spherical to sub-spherical, light brown pycnidia, with a marked opening or ostiole through which the conidia are released
spores	oval to ellipsoidal, colourless, one-celled

DIFFERENTIAL DIAGNOSIS

colonial appearance   *Lecythophora hoffmannii*

microscopic appearance   other *Phoma* spp., though these often have black pycnidia

SEXUAL STATE

None known.

CLINICAL IMPORTANCE

It is a rare cause of human infection.

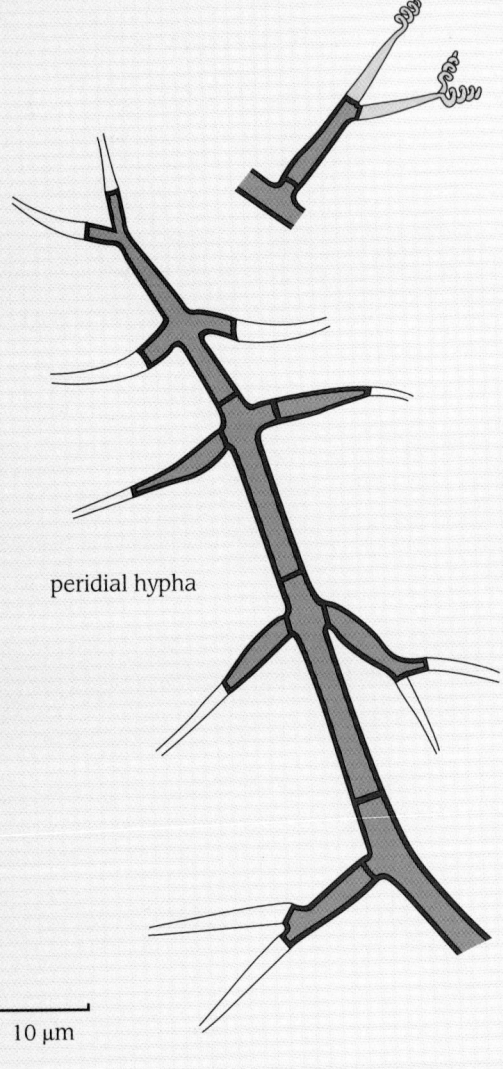
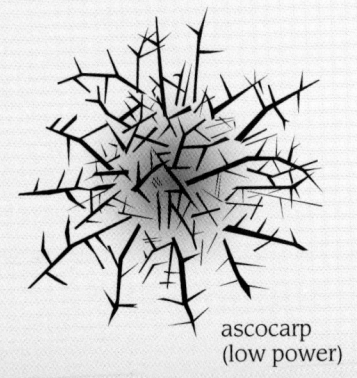

MYXOTRICHUM DEFLEXUM

COLONIAL APPEARANCE
at 30°C on glucose peptone agar

diameter	15 mm in one week
topography	flat
texture	floccose, with secreted droplets
colour	greenish-grey
reverse	red, with a diffusing, bright orange-red pigment

MICROSCOPIC APPEARANCE
at 30°C

predominant features	large, dark-brown balls of branching hyphae (non-walled ascocarps)
fruiting bodies	ascocarps composed of an interwoven mass of peridial hyphae with thorn-like branches each terminating in minute, spiral hyphae; eight-spored, club-shaped asci develop within
ascospores	oval to ellipsoidal, 3.5–5.5 µm × 3 µm, colourless to pale yellow, with longitudinal striations

DIFFERENTIAL DIAGNOSIS

colonial appearance   *Penicillium marneffei* has a similar diffusing, red pigment

microscopic appearance   other members of the *Gymnoascacae* have non-walled ascocarps, but with peridial hyphae of different form from those of *Myxotrichum deflexum*

CLINICAL IMPORTANCE

This keratinophilic soil fungus is a possible cause of onychomycosis.

SCHIZOPHYLLUM COMMUNE

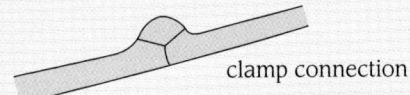

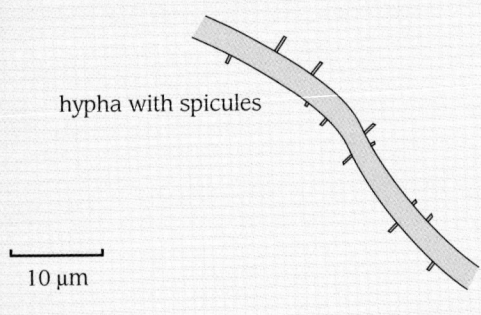

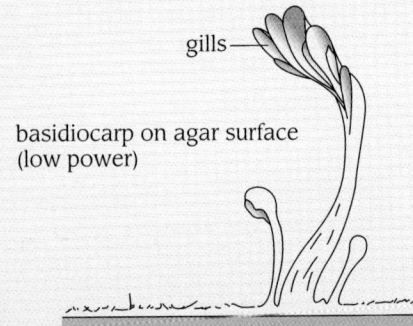

COLONIAL APPEARANCE
at 30°C on glucose peptone agar

diameter	70 mm in one week
topography	flat; produces raised fruiting bodies on prolonged incubation
texture	floccose
colour	off-white to pale brown
reverse	cream

MICROSCOPIC APPEARANCE
at 30°C

predominant features	colourless, septate mycelium, often with clamp connections; minute spicules may project from the hyphal walls
fruiting bodies	kidney-shaped or 'mushroom-like' fruiting bodies, several centimetres across, formed in concentric zones; gills formed on the lower side from basidia, each with four projections supporting a basidiospore
basidiospores	colourless, oval to pear-shaped, 7 μm × 3 μm

DIFFERENTIAL DIAGNOSIS

colonial appearance — other saprophytic Basidiomycetes produce similar, fast-growing, white, floccose colonies

microscopic appearance — other saprophytic Basidiomycetes have hyphae with clamp connections but lack hyphal spicules, and have different fruiting bodies

CLINICAL IMPORTANCE

This Basidiomycete is often found on rotten wood. It is a rare cause of human disease, but sinusitis and endocarditis are among the deep infections that have been reported.

LEPTOSPHAERIA SENEGALENSIS

COLONIAL APPEARANCE
at 30°C on glucose peptone agar

diameter	10 mm in one week
topography	flat
texture	floccose
colour	dark olive, with grey margin
reverse	dark brown to black

MICROSCOPIC APPEARANCE
at 30°C

predominant features	brown hyphae only on glucose peptone agar; ascocarps formed after one month on cornmeal agar
fruiting bodies	large, black, spherical ascocarps, 100-300 µm in diameter, submerged in the agar; ascocarp walls composed of thick-walled cells; long, club-shaped asci, each with eight ascospores; asci surrounded by unbranched hyphae (paraphyses)
ascospores	oval, usually with four septa, 23-30 µm × 8-10 µm, with septal constrictions

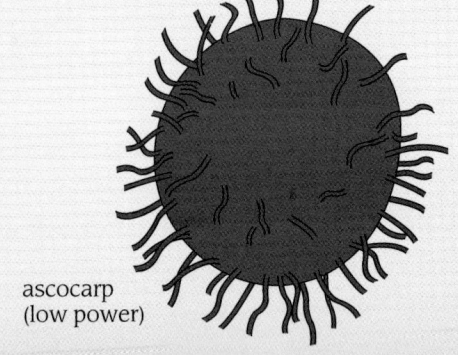

DIFFERENTIAL DIAGNOSIS

colonial appearance   *Phialophora* spp., *Cladosporium* spp., *Exophiala* spp.

microscopic appearance   many fungi produce submerged, black ascocarps with septate ascospores; referral to a specialist is recommended

CLINICAL IMPORTANCE

It is a cause of black-grain eumycetoma in Africa.

NEOTESTUDINA ROSATII

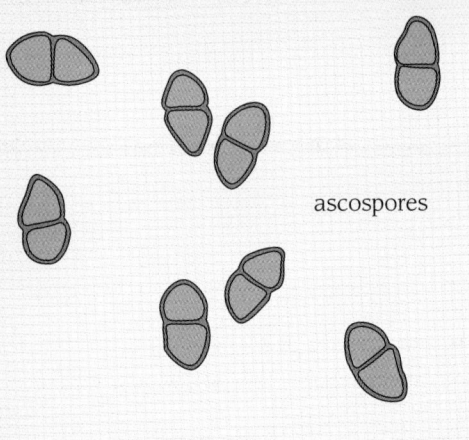

ascospores

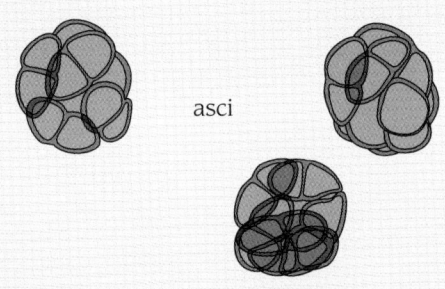

asci

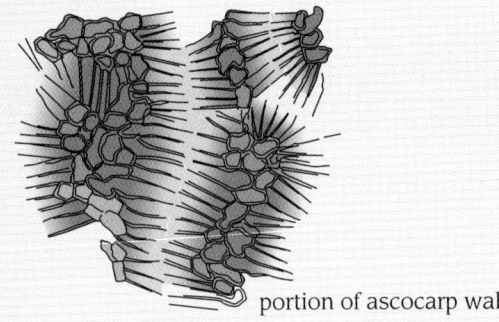

portion of ascocarp wall

10 μm

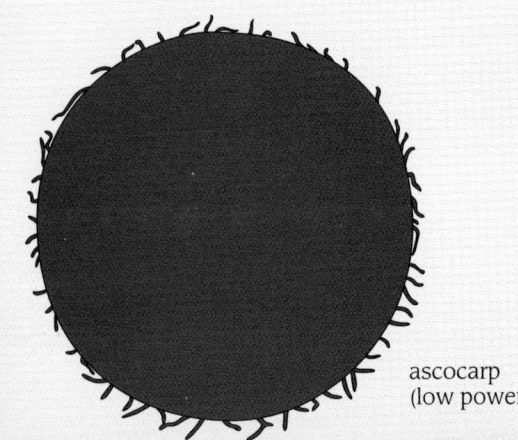

ascocarp
(low power)

COLONIAL APPEARANCE
at 30°C on glucose peptone agar

diameter	10 mm in one week
topography	flat and grooved or folded
texture	leathery, with short aerial mycelium
colour	brown to dark grey
reverse	dark brown

MICROSCOPIC APPEARANCE
at 30°C

predominant features	brown hyphae only on glucose peptone agar; black ascocarps formed after one month on cornmeal agar
fruiting bodies	large, black, spherical ascocarps, 180-900 μm in diameter, submerged in the agar; ascocarp walls composed of plates of irregularly-shaped cells; club-shaped to round asci, each containing eight ascospores
ascospores	thick-walled, brown, rhomboidal, 9-12.5 μm × 4.5-8 μm, with one septum

DIFFERENTIAL DIAGNOSIS

colonial appearance   *Madurella* spp., *Exophiala* spp., *Phialophora* spp.

microscopic appearance   other non-sporing, brown fungi such as *Madurella* spp.; the ascocarp state should be referred to a specialist for identification

CLINICAL IMPORTANCE

It is a cause of pale-grain eumycetoma.

PIEDRAIA HORTAE

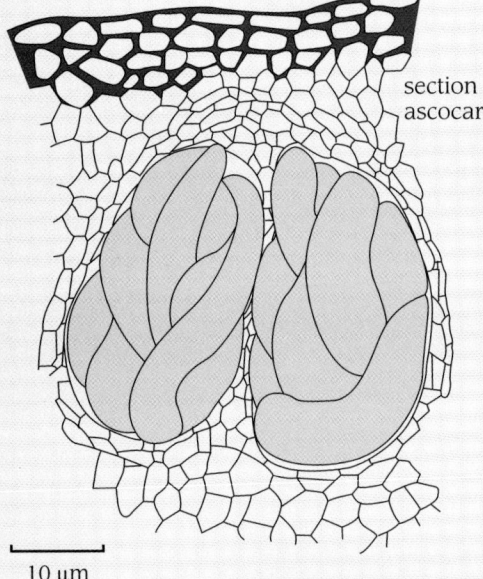

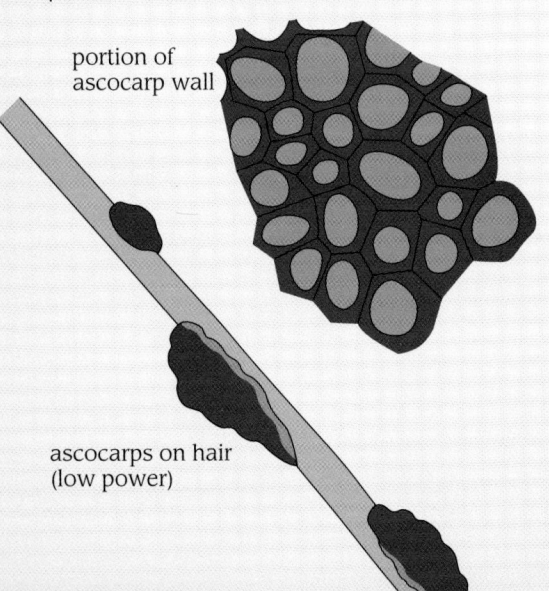

COLONIAL APPEARANCE
at 30°C on glucose peptone agar

diameter	10 mm in one week
topography	conical, folded with flat margin
texture	glabrous, becoming covered with short, aerial mycelium
colour	dark brown to black
reverse	dark brown to black; sometimes with a diffusing, brown pigment

MICROSCOPIC APPEARANCE
at 30°C

predominant features	thick-walled, septate, pigmented hyphae with intercalary chlamydospores; black ascocarps seldom produced on glucose peptone agar
fruiting bodies	ascocarps variable in size and shape, with an irregular surface; ascocarps usually contain a few ellipsoidal asci, each with eight ascospores
ascospores	colourless, cylindrical, non-septate, with curved ends tapering into hair-like projections

DIFFERENTIAL DIAGNOSIS

colonial appearance   other moulds with dark brown colonies

microscopic appearance   *Madurella* spp. and other non-sporing black moulds; the ascocarp state should be referred to a specialist for identification if clinical details suggests an infection other than black piedra

CLINICAL IMPORTANCE

It is the cause of black piedra, an infection of the scalp hair that occurs in tropical regions.

LASIODIPLODIA THEOBROMAE

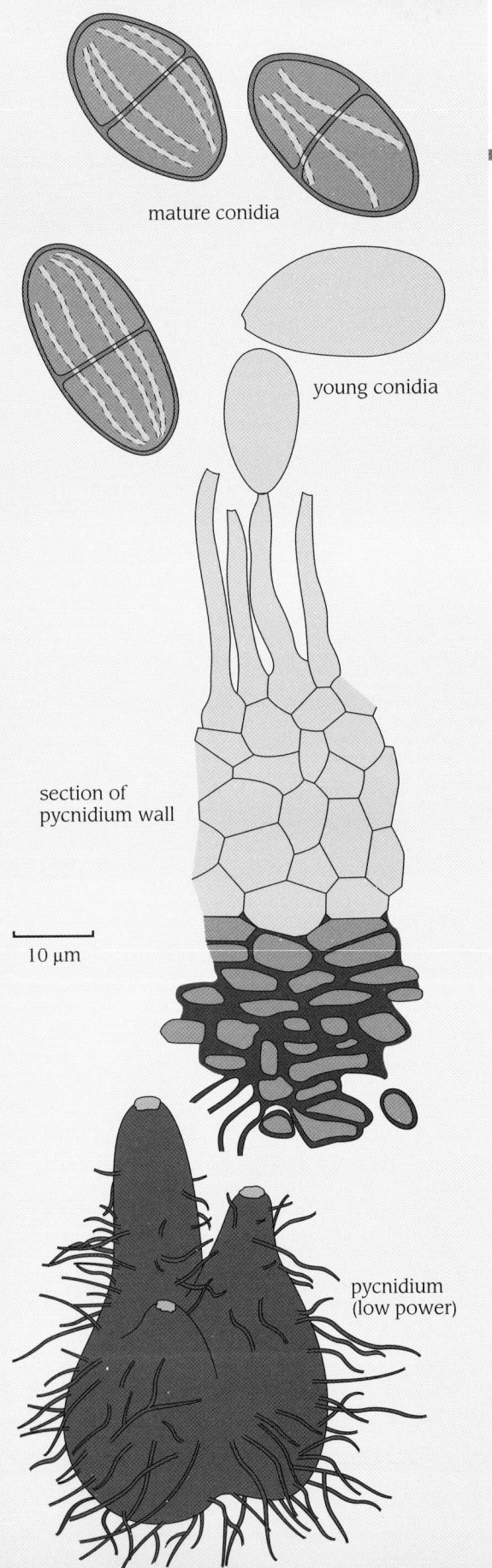

COLONIAL APPEARANCE
at 30°C on glucose peptone agar

diameter	90 mm in one week
topography	abundant aerial growth to lid of petri dish
texture	floccose
colour	grey to brown-black
reverse	black

MICROSCOPIC APPEARANCE
at 30°C

predominant features	brown hyphae only on glucose peptone agar; pycnidial stromata stimulated on cornmeal or other poor media
fruiting bodies	large, black, flask-shaped pycnidial stromata, up to 5 mm in diameter; made up of several pycnidia, each with a wide ostiole that releases many colourless or dark brown conidia
spores	colourless, ellipsoidal, non-septate conidia, 20-30 µm × 10-15 µm which become dark brown and septate, with longitudinal striations and truncate bases

DIFFERENTIAL DIAGNOSIS

colonial appearance *Scytalidium dimidiatum*

microscopic appearance other moulds with septate pycnidiospores

SEXUAL STATE

None known.

CLINICAL IMPORTANCE

It is a rare cause of corneal and nail infections.

PYRENOCHAETA ROMEROI

conidia

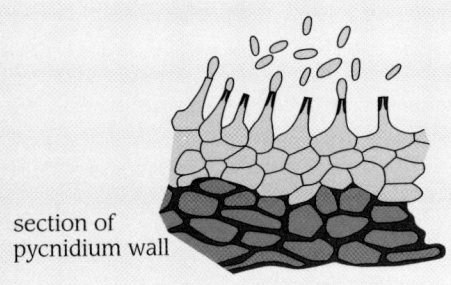
section of pycnidium wall

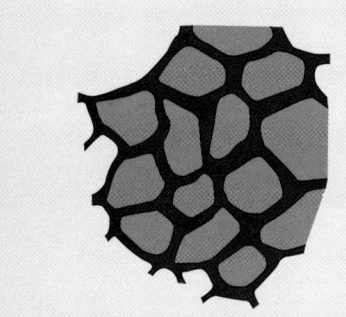
portion of pycnidium wall

10 μm

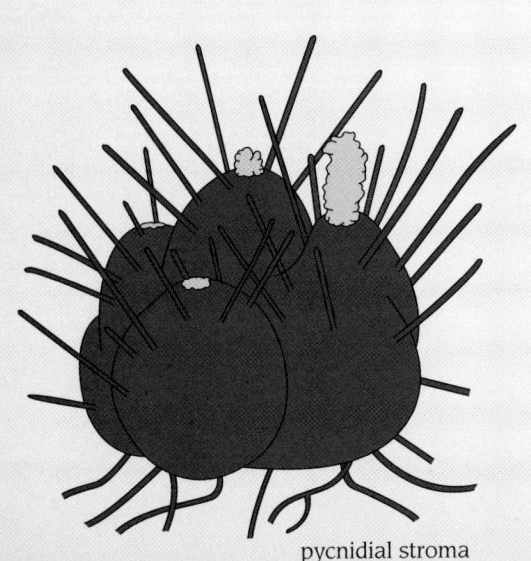
pycnidial stroma (low power)

COLONIAL APPEARANCE
at 30°C on glucose peptone agar

diameter	10 mm in one week
topography	flat to domed
texture	densely velvety to floccose
colour	silvery grey
reverse	olive-black

MICROSCOPIC APPEARANCE
at 30°C

predominant features	brown, septate mycelium; brown-black pycnidia develop after several weeks
fruiting bodies	oval to cylindrical, brown-black pycnidia, 80-150 μm, with dark hyphae projecting like spines from the thick, multilayered pycnidial wall
spores	colourless, ellipsoidal conidia, 2 μm × 1 μm; formed from short, flask-shaped phialides lining the inner pycnidial wall and emerging through the ostiole in slimy drops

DIFFERENTIAL DIAGNOSIS

colonial appearance
: *Madurella grisea* is identical and may be a non-pycnidial form of *Pyrenochaeta romeroi*; some *Phialophora* spp. and *Cladophialophora* spp. are similar

microscopic appearance
: *Phoma* spp., but these do not have the spine-like hyphae on the pycnidia

SEXUAL STATE

None known.

CLINICAL IMPORTANCE

It causes dark-grain eumycetoma in South America.

PYRENOCHAETA UNGUIS-HOMINIS

This species is similar, but the pycnidia are lined by longer, branching conidiophores. It is a rare cause of onychomycosis.

MADURELLA MYCETOMATIS

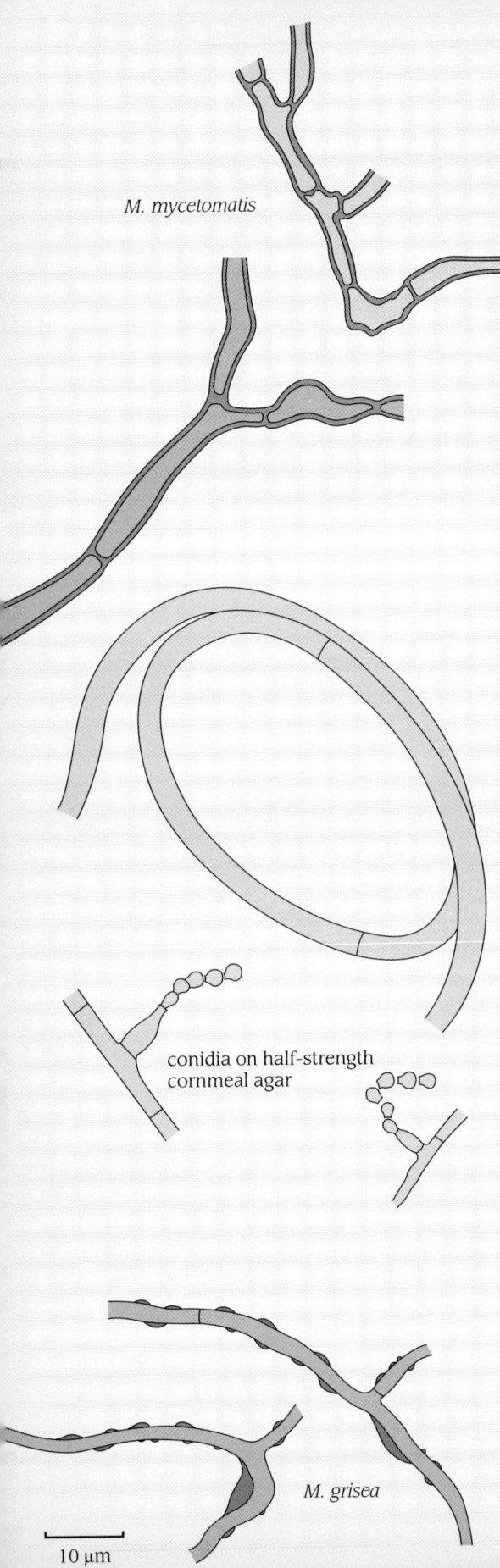

COLONIAL APPEARANCE
at 30°C on glucose peptone agar

diameter	20 mm in one week
topography	flat, with raised centre and radiating folds
texture	velvety
colour	cream, yellow or olive brown
reverse	pale, sometimes with a diffusing, brown pigment

MICROSCOPIC APPEARANCE
at 30°C

predominant features	usually sterile, pigmented mycelium with chlamydospores; phialides may be induced on special media (half-strength cornmeal agar)
conidia	small, oval to round, produced from tips of mostly lateral phialides

DIFFERENTIAL DIAGNOSIS

colonial appearance   *Madurella grisea*; other moulds with brown colonies

microscopic appearance   *M. grisea* and many other brown, non-sporing moulds

Note: a clinical and histological diagnosis of black-grain eumycetoma is required for definitive identification.

SEXUAL STATE

None known.

CLINICAL IMPORTANCE

It is a common cause of black-grain eumycetoma in tropical and sub-tropical regions.

MADURELLA GRISEA

This non-sporing organism is identical in colonial appearance with *Pyrenochaeta romeroi* and may represent a non-pycnidial form of the same fungus. It can be differentiated from *M. mycetomatis* by its colonial appearance and by its optimal growth temperature of 30°C (the optimum for *M. mycetomatis* is 37°C). It is a cause of black-grain eumycetoma in Central and South America.

11 IDENTIFICATION OF YEASTS

INTRODUCTION

A combination of morphological and biochemical tests are required to identify a yeast. Useful morphological characteristics include the colour of the colonies, the size and shape of the cells, the presence of a capsule, the production of hyphae or pseudohyphae, and the production of chlamydospores. Useful biochemical tests include the assimilation and fermentation of sugars, and the assimilation of nitrate. Most yeasts associated with human infections can be identified using one of the commercial test products that are based on sugar assimilation of isolates. However, it is important to remember that morphological examination is essential to avoid confusion between organisms with identical biochemical profiles. In addition there are a number of simple tests for the presumptive identification of some of the most important yeasts. These include the serum germ tube test for the rapid identification of *Candida albicans* and the urease test for *Cryptococcus neoformans*.

Yeasts are unicellular fungi which consist of round, oval or elongated cells, or *blastospores*, that propagate by budding out similar cells from their surfaces. The bud may become detached from the parent cell, or remain attached and produce another bud itself. In this way a chain of cells may be produced. The term *pseudohypha* is used to describe a chain of yeast cells which have become elongated before budding and have remained attached to one another. Unlike a true hypha, the connection between adjacent pseudohyphal cells shows a marked constriction. In addition to pseudohyphae,

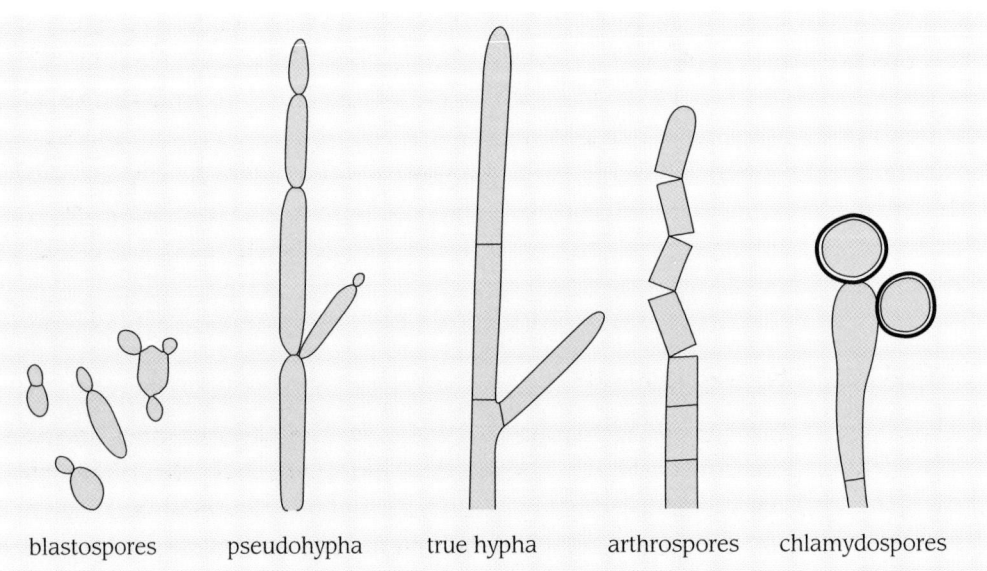

some species of *Candida* and *Trichosporon* can produce true hyphae. The hyphae of these organisms, like those of the moulds described in Chapter 3, can fragment to form chains of individual cells, or *arthrospores*. Although it is an arthrosporic mould, *Geotrichum candidum* has been included in this chapter because its colonies are similar in appearance to those of *Blastoschizomyces capitatus* and *Trichosporon beigelii*.

There are two simple tests that should be performed on yeast isolates: the germ tube test and microscopic examination for the presence of a capsule. If the germ tube test is positive, the organism is *Candida albicans*. If a capsule is present, this permits the presumptive identification of the organism as *Cryptococcus neoformans*, but further tests are necessary to confirm it. If germ tubes are not produced and a capsule is absent, further morphological and biochemical tests should be performed.

GERM TUBE TEST

The germ tube test allows a rapid identification of *Candida albicans* using either the original isolation plate or a purified culture. The test consists in taking a light loopful of inoculum from a culture plate, suspending it in 0.5 mL of sterile horse serum and incubating at 37°C for 2-3 h. A drop of the suspension is then placed on a microscope slide with a cover slip and the preparation examined under a microscope. The isolate under test is *C. albicans* if the cells have produced short hyphae and there is no constriction at the junction between the parent cell and the hypha.

This test has some drawbacks. Fewer than 10% of blastospores produce germ tubes in many positive tests; about 5% of *Candida albicans* isolates fail to produce germ tubes; over-inoculation of the serum can result in inhibition of germ tube formation; and too short an incubation period can lead to false negative results: 2 h is a minimum, not a maximum incubation time. Confusion often arises between *C. albicans* and *C. tropicalis* in the germ tube test, as the yeast cells of the latter tend to produce pseudohyphae. The distinction is made by observing the constriction between the parent cell and the pseudohypha in *C. tropicalis*.

CAPSULE PRODUCTION

Cryptococcus neoformans produces round to oval cells with polysaccharide capsules. These can be detected when the cells are mounted in a pigmented colloidal mounting fluid, such as India ink, which does not penetrate the capsular envelope. The test consists in taking a light loopful of inoculum from the culture and suspending it in a drop of 50% aqueous India ink on a microscope slide. If a capsule is present, it should be visible as a clear halo around the cells. The presence of a capsule gives a presumptive identification of *Cr. neoformans*, but does not provide a definitive identification. Drawbacks of this test include lack of prominent capsules in some *Cr. neoformans* isolates and the possibility of loss of the capsule following subculture of the organism.

UREASE TEST

Urease production is a characteristic of most *Cryptococcus neoformans* isolates and is useful for their presumptive identification. The test consists in taking a light loopful of inoculum from the original isolation plate, spreading it over the surface of a Christensen's urea slope and incubating at 30°C for up to four days. A colour change from amber to pink permits the presumptive identification of the isolate as *Cr. neoformans*. However, other species of *Cryptococcus*, as well as *Rhodotorula* and *Trichosporon* species, can give a positive result. Bacterial contamination can also result in a change in the colour of the medium.

MORPHOLOGICAL EXAMINATION

Morphological examination of yeast isolates under the microscope is essential to avoid errors in identification of organisms with identical biochemical profiles. Growth in microaerophilic conditions on cornmeal or other starch-containing media, such as rice agar, stimulates the formation of hyphae, pseudohyphae, arthrospores and chlamydospores in those species able to produce them.

The surface of the medium (cornmeal agar) is inoculated across the centre of the plate using a wire loop. A sterile cover slip is placed over part of the inoculum and the plate incubated at 30°C for at least 48 h. At intervals, for a period of up to one week, the lid of the plate is removed and the growth under the cover slip examined under the low power objective of a microscope.

Chlamydospores (which can take up to four days to develop on cornmeal agar) are indicative of *Candida albicans*. Use of Czapek Dox plus Tween 80 medium often stimulates their production within 24 h. If only pseudohyphae are found, the isolate is a *Candida* species other than *C. albicans*. If arthrospores are present, the isolate is a presumptive *Trichosporon* species. However, it should be noted that *C. guilliermondii* needs three to four days' incubation to produce pseudohyphae while *T. beigelii* needs a similar time to produce arthrospores.

BIOCHEMICAL TESTS

Most laboratories now use commercial identification systems to determine the biochemical profile of yeast isolates. These kits are less time-consuming to set up, simpler to interpret, and often permit more rapid identification of isolates than the classical assimilation and fermentation methods which they have replaced. Some offer extensive databases capable of identifying a wide range of organisms, but others are much more limited in scope.

Information on the full biochemical profiles of the organisms described in this chapter will be found in most commercial identification systems. For convenience, the principal assimilation and fermentation reactions used in comparing organisms are listed here in Tables 11.1 and 11.2.

Table 11.1 Assimilation reactions

	GLU	GAL	SUC	MAL	LAC	RAF	CEL	RHA	TRE
B. capitatus	+	+	−	−	−	−	−	−	−
C. albicans	+	+	+	+	−	−	−	−	+
C. glabrata	+	−	−	−	−	−	−	−	+
C. guilliermondii	+	+	+	+	−	+	+	V	+
C. kefyr	+	+	+	−	+	+	V	−	V
C. krusei	+	−	−	−	−	−	−	−	−
C. lipolytica	+	−	−	−	−	−	−	−	−
C. lusitaniae	+	V	+	+	−	−	+	+	+
C. parapsilosis	+	+	+	+	−	−	−	−	+
C. pelliculosa	+	V	+	+	−	V	+	−	+
C. tropicalis	+	+	V	+	−	−	V	−	+
Cr. neoformans	+	+	+	+	−	+	V	+	+
G. candidum	+	+	−	−	−	−	−	−	−
R. glutinis	+	+	+	+	−	+	+	V	+
S. cerevisiae	+	+	+	+	−	+	−	−	V
T. beigelii	+	V	V	V	+	V	V	V	V

Key: CEL, cellobiose; GAL, galactose; GLU, glucose; LAC, lactose; MAL, maltose; RAF, raffinose; RHA, rhamnose; SUC, sucrose; TRE, trehalose; V, variable reaction

Table 11.2 Fermentation and other reactions

	GLU	SUC	MAL	LAC	TRE	NIT	URE
B. capitatus	−	−	−	−	−	−	−
C. albicans	+	−	+	−	+	−	−
C. glabrata	+	−	−	−	+	−	−
C. guilliermondii	+	+	−	−	+	−	−
C. kefyr	+	+	−	+	−	−	−
C. krusei	+	−	−	−	−	−	−
C. lipolytica	−	−	−	−	−	−	+
C. lusitaniae	+	+	V	−	+	−	−
C. parapsilosis	+	−	−	−	−	−	−
C. pelliculosa	+	+	V	−	−	+	−
C. tropicalis	+	+	+	−	+	−	−
Cr. neoformans	−	−	−	−	−	−	+
G. candidum	−	−	−	−	−	−	−
R. glutinis	−	−	−	−	−	+	+
S. cerevisiae	+	+	+	−	V	−	−
T. beigelii	−	−	−	−	−	−	+

Key: GLU, glucose fermentation; SUC, sucrose fermentation; MAL, maltose fermentation; LAC, lactose fermentation; TRE, trehalose fermentation; NIT, nitrate assimilation; URE, urease production; V, variable reaction

Unlike earlier chapters, the descriptions in this chapter do not follow the order of the key. As in earlier chapters, the descriptions for organisms of similar morphological appearance will be found on adjacent pages.

MEDIA FOR YEAST IDENTIFICATION

Christensen's urea agar

This medium is useful for the presumptive identification of *Cryptococcus neoformans*. It is important to remember that other species of *Cryptococcus*, as well as *Rhodotorula* and *Trichosporon* species, can also give a positive result.

glucose	1 g
mycological peptone	1 g
sodium chloride	5 g
potassium dihydrogen orthophosphate	2 g
phenol red	0.012 g
agar	15 g
distilled water	1 L

Heat to dissolve. Autoclave at 115°C for 20 min. Cool to 50°C and add 50 mL of sterile 40% urea solution.

Cornmeal agar

This medium is useful for stimulating the formation of pseudohyphae, true hyphae, arthrospores and chlamydospores in those species able to produce them.

cornmeal extract	2 g
agar	15 g
distilled water	1 L

Heat to dissolve. Autoclave at 121°C for 15 min.

Czapek-Dox plus Tween 80 agar

This medium is useful for stimulating chlamydospore production in *Candida albicans*.

sucrose	30 g
sodium nitrate	2 g
potassium chloride	0.5 g
magnesium glycerophosphate	0.5 g
potassium sulphate	0.35 g
ferrous sulphate	0.01 g
agar	12 g
Tween 80	10 mL
distilled water	1 L

Heat to dissolve. Autoclave at 121°C for 15 min.

Sabouraud's glucose peptone agar This medium is recommended for the isolation and cultivation of yeasts. Antibacterial antibiotics (in particular chloramphenicol) can be added to control bacterial contamination.

glucose	40 g
mycological peptone	10 g
agar	15 g
distilled water	1 L

Heat to dissolve. Autoclave at 121°C for 15 min.

Key to yeasts

1a	Minute colonies on glucose peptone agar	presumptive *Malassezia furfur*
1b	Pink or red colonies on glucose peptoneagar	presumptive *Rhodotorula* sp. (determine biochemical profile)
1c	White or cream colonies on glucose peptone agar	2
2a	Germ tube test positive	*Candida albicans*
2b	Germ tube test negative	3
3a	Capsule present	presumptive *Cryptococcus neoformans* (determine biochemical profile)
3b	Capsule absent	4
4a	Urease test positive	presumptive *Cryptococcus neoformans* or *Trichosporon beigelii* or *Malassezia pachydermatis* (determine biochemical profile)
4b	Urease test negative	5
5a	Chlamydospores present on cornmeal agar	*Candida albicans*
5b	No chlamydospores on cornmeal agar	6
6a	Arthrospores present on cornmeal agar	7
6b	No arthrospores on cornmeal agar	8
7a	Budding cells present on cornmeal agar	presumptive *Trichosporon* sp. or *Blastoschizomyces capitatus* (determine biochemical profile)
7b	No budding cells on cornmeal agar; dichotomously branching hyphae present	*Geotrichum candidum*
8a	Pseudohyphae present on cornmeal agar	presumptive *Candida* sp. (determine biochemical profile)
8b	No pseudohyphae on cornmeal agar	determine biochemical profile

Note: many *Candida* species do not form pseudohyphae on cornmeal agar. These organisms were formerly classified in the genus *Torulopsis*.

CANDIDA ALBICANS

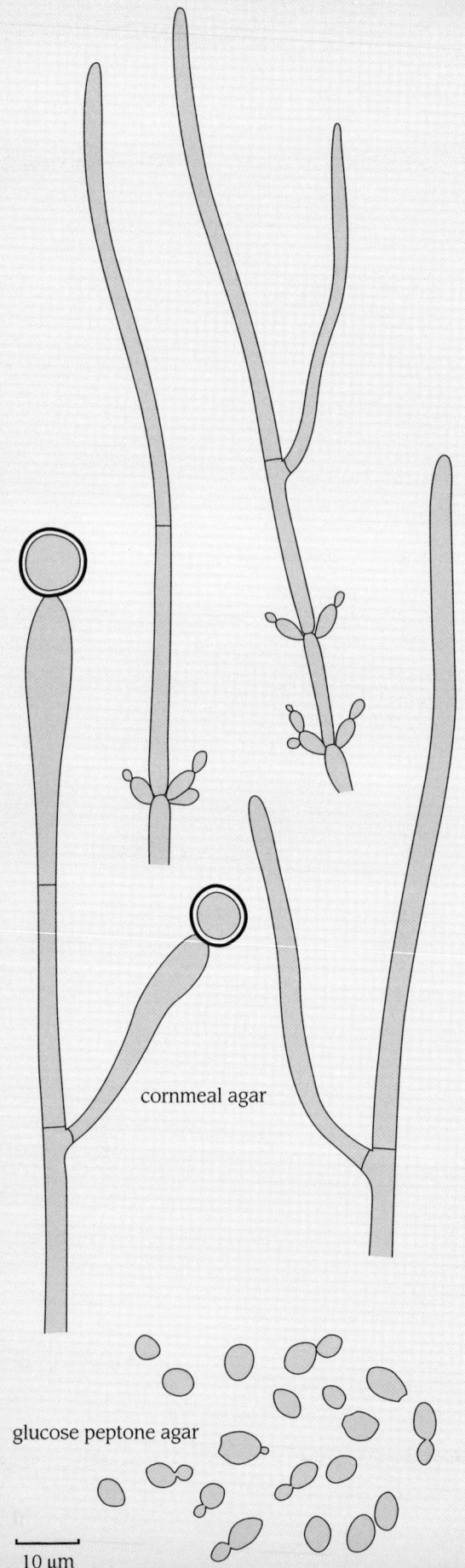

COLONIAL APPEARANCE
at 30°C on glucose peptone agar

colour	white to cream
surface characteristics	glistening, smooth; sometimes dull, rough; with or without a filamentous margin

MICROSCOPIC APPEARANCE
at 30°C on cornmeal agar

predominant features	true hyphae, pseudohyphae and chlamydospores; blastospores formed in clusters at intervals along the length of the hyphae

DIFFERENTIAL DIAGNOSIS

Candida dubliniensis (not described) has an identical biochemical profile and morphological appearance, but chlamydospore production is much more abundant.

C. tropicalis has a similar morphological appearance but is able to ferment sucrose.

SEXUAL STATE

None known.

CLINICAL IMPORTANCE

It can cause localised or disseminated deep-seated infection in debilitated or immunocompromised individuals, but is more commonly seen causing mucosal, cutaneous or nail infection. It can be isolated from the mouth and gastrointestinal tract of 30-50% of the normal population. Most human infections are endogenous in origin. The name *C. dubliniensis* has been applied to chlamydosporic strains recovered from patients with HIV infection.

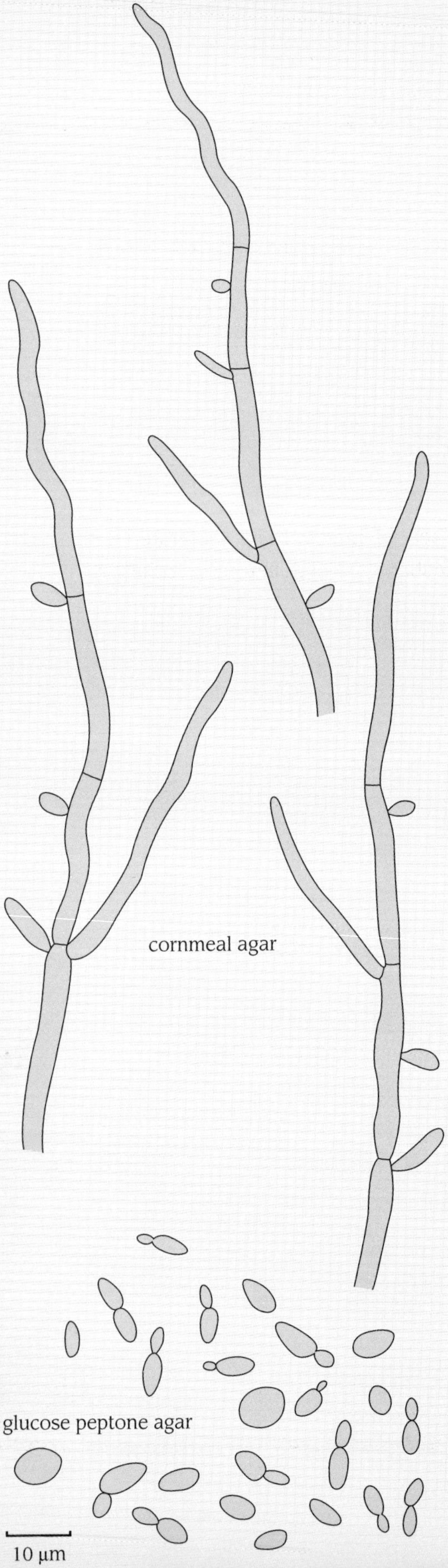

CANDIDA TROPICALIS

COLONIAL APPEARANCE

at 30°C on glucose peptone agar

colour	white to cream
surface characteristics	glistening, smooth; sometimes dull, rough

MICROSCOPIC APPEARANCE

at 30°C on cornmeal agar

predominant features	true hyphae and pseudohyphae; blastospores formed in clusters at intervals along the hyphae; single internodal blastospores sometimes found

DIFFERENTIAL DIAGNOSIS

Candida albicans has a similar morphological appearance, but forms chlamydospores, lacks internodal blastospores, and is unable to ferment sucrose.

SEXUAL STATE

None known.

CLINICAL IMPORTANCE

It is an important cause of mucosal and deep-seated infection in debilitated or immunocompromised individuals. Most human infections are endogenous in origin.

CANDIDA KRUSEI

COLONIAL APPEARANCE
at 30°C on glucose peptone agar

colour	white to grey
surface characteristics	flat, dull, smooth

MICROSCOPIC APPEARANCE
at 30°C on cornmeal agar

predominant features	abundant pseudohyphae; minimal constriction at the base of the terminal pseudohypha giving the appearance of a true hypha; ellipsoidal to cylindrical blastospores

DIFFERENTIAL DIAGNOSIS

Candida kefyr has elongated blastospores but a different biochemical profile.

SEXUAL STATE

Issatchenkia orientalis

CLINICAL IMPORTANCE

It is an occasional cause of mucosal and deep-seated infection in debilitated or immunocompromised individuals. Its emergence as a significant pathogen has been associated with increasing use of fluconazole, to which it is resistant.

CANDIDA LIPOLYTICA

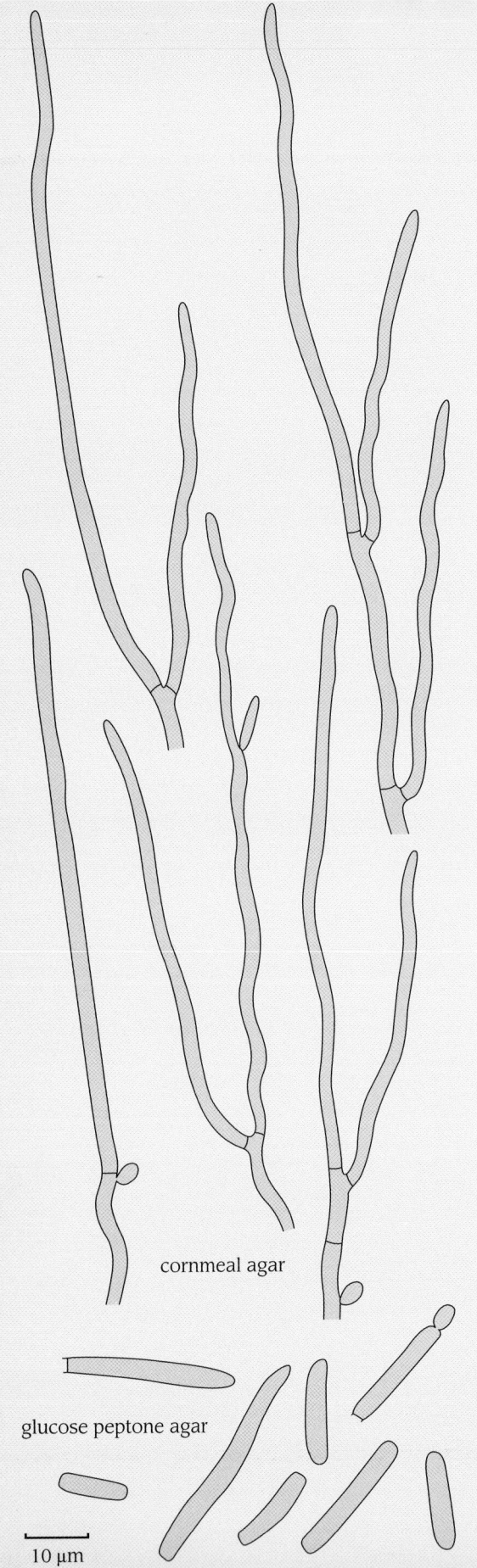

COLONIAL APPEARANCE
at 30°C on glucose peptone agar

colour	cream
surface characteristics	raised, dull

MICROSCOPIC APPEARANCE
at 30°C on cornmeal agar

predominant features	true hyphae and pseudohyphae; single or paired, nodal or internodal blastospores

DIFFERENTIAL DIAGNOSIS

Candida albicans and *C. tropicalis* also form true hyphae but have a different biochemical profile and do not produce urease.

SEXUAL STATE

Yarrowia lipolytica

CLINICAL IMPORTANCE

It is a rare cause of human infection.

CANDIDA KEFYR

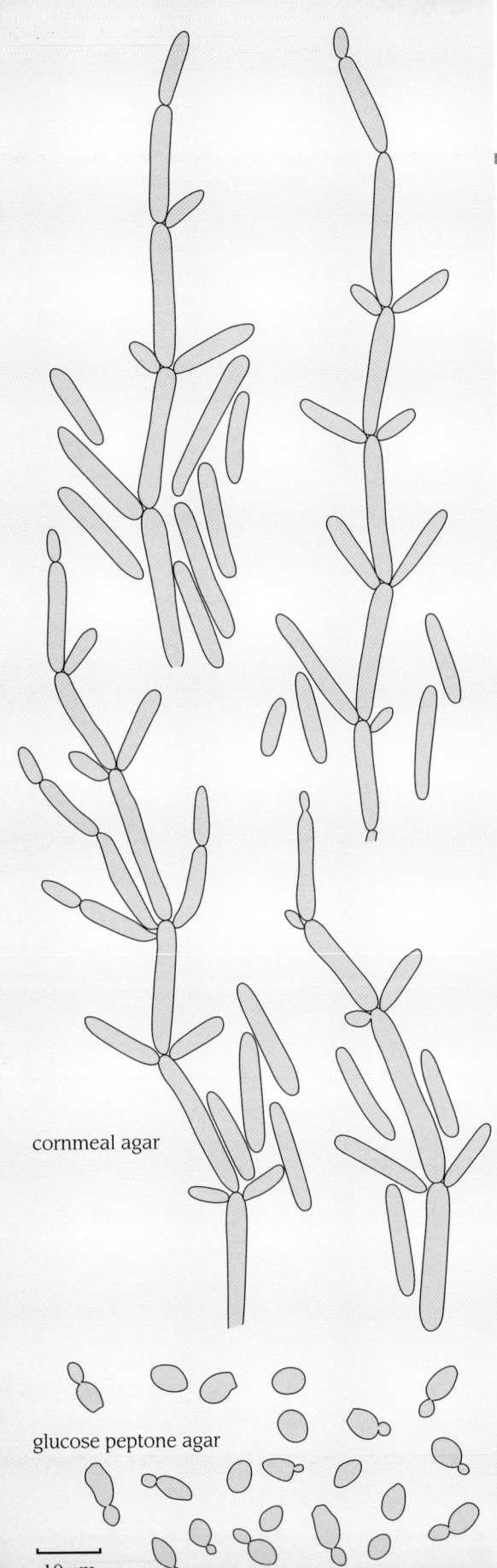

cornmeal agar

glucose peptone agar

10 μm

COLONIAL APPEARANCE
at 30°C on glucose peptone agar

colour cream

surface
characteristics dull, smooth

MICROSCOPIC APPEARANCE
at 30°C on cornmeal agar

predominant features large clusters of detached pseudohyphal cells; elongated blastospores

DIFFERENTIAL DIAGNOSIS

Candida krusei has similar elongated blastospores but a different biochemical profile.

SEXUAL STATE

Kluyveromyces marxianus

CLINICAL IMPORTANCE

It is an occasional cause of human infection.

CANDIDA LUSITANIAE

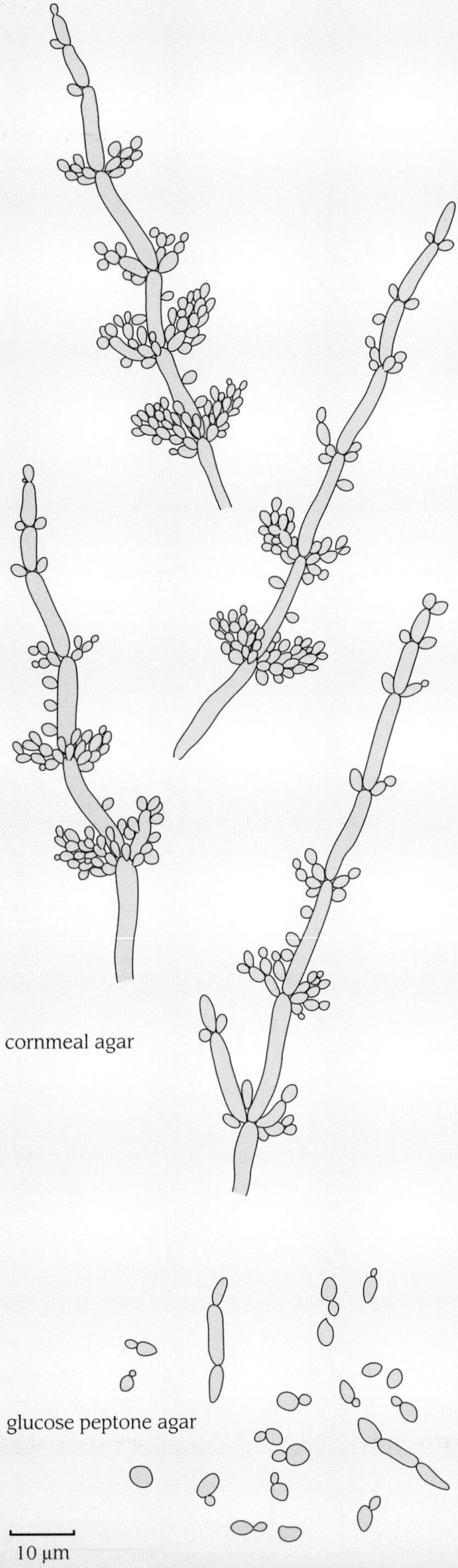

cornmeal agar

glucose peptone agar

10 µm

COLONIAL APPEARANCE
at 30°C on glucose peptone agar

colour white to cream

surface characteristics glistening, smooth

MICROSCOPIC APPEARANCE
at 30°C on cornmeal agar

predominant features long pseudohyphae with few branches; abundant, small, oval blastospores

DIFFERENTIAL DIAGNOSIS

Candida kefyr, *C. parapsilosis* and *C. guilliermondii* can have a similar morphological appearance to *C. lusitaniae*, but have a different biochemical profile and in particular do not assimilate rhamnose.

SEXUAL STATE

Clavispora lusitaniae

CLINICAL IMPORTANCE

It is an occasional cause of deep-seated infection in debilitated or immunocompromised patients. Its emergence as a significant pathogen has been associated with use of amphotericin B, to which it is often resistant.

CANDIDA PARAPSILOSIS

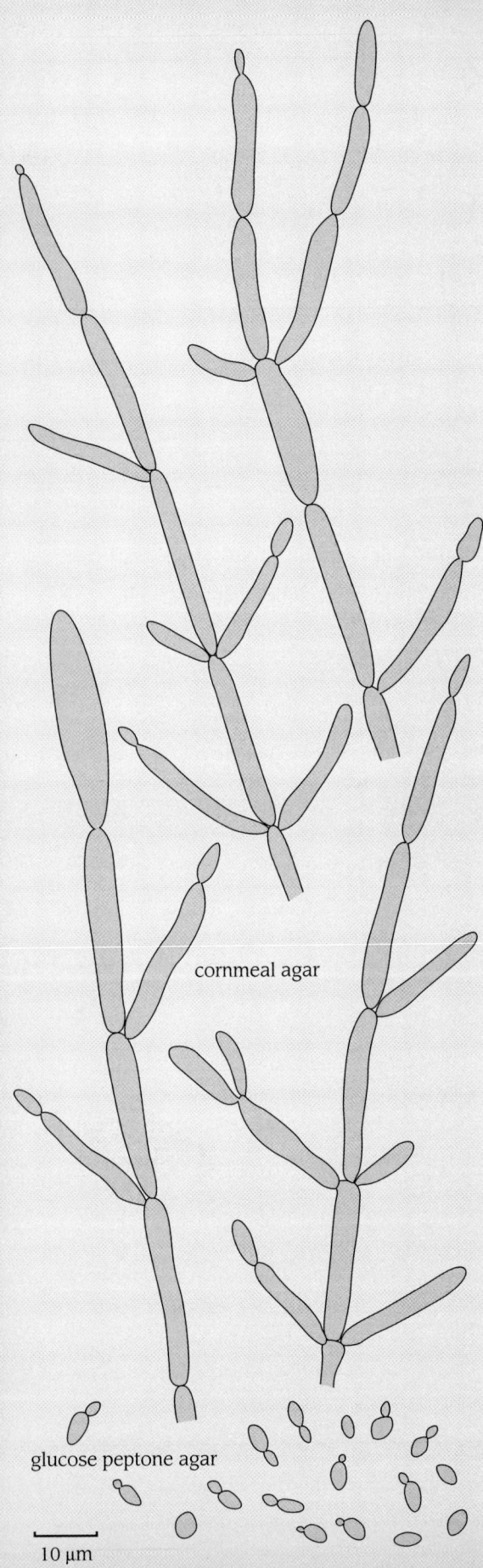

COLONIAL APPEARANCE
at 30°C on glucose peptone agar

colour	cream
surface characteristics	glistening, smooth; or dull, rough

MICROSCOPIC APPEARANCE
at 30°C on cornmeal agar

predominant features	branching pseudohyphae; some pseudohyphae terminate in a swollen cell; occasional round to oval blastospores

DIFFERENTIAL DIAGNOSIS

Candida kefyr and *C. guilliermondii* are similar in morphological appearance to *C. parapsilosis* but have a different biochemical profile.

SEXUAL STATE

None known.

CLINICAL IMPORTANCE

It is a common skin commensal and an occasional cause of superficial candidosis. It has also caused deep-seated infection, often in association with parenteral nutrition.

CANDIDA PELLICULOSA

cornmeal agar

glucose peptone agar

10 µm

COLONIAL APPEARANCE
at 30°C on glucose peptone agar

colour	white to cream
surface characteristics	glistening, smooth; or dull, rough

MICROSCOPIC APPEARANCE
at 30°C on cornmeal agar

predominant features	branching pseudohyphae; round to ellipsoidal blastospores

DIFFERENTIAL DIAGNOSIS

Other *Candida* spp. can have a similar morphological appearance, but have a different biochemical profile and do not assimilate nitrate.

SEXUAL STATE

Pichia anomala

CLINICAL IMPORTANCE

It is an occasional cause of deep-seated infection in debilitated or immunocompromised individuals, often in association with catheterisation.

CANDIDA GLABRATA

edge of streak on cornmeal agar

glucose peptone agar

10 µm

COLONIAL APPEARANCE
at 30°C on glucose peptone agar

colour	cream
surface characteristics	glistening, smooth

MICROSCOPIC APPEARANCE
at 30°C on cornmeal agar

predominant features	small, round to oval blastospores

DIFFERENTIAL DIAGNOSIS

Candida famata (not described), *C. inconspicua* (not described), and *Saccharomyces cerevisiae* can have a similar morphological appearance but have larger cells and a different biochemical profile.

SEXUAL STATE

None known.

CLINICAL IMPORTANCE

It is a cause of mucosal and deep-seated infection. Its emergence as a significant pathogen has been associated with increasing use of azole compounds, to which it is often resistant.

CANDIDA GUILLIERMONDII

cornmeal agar

glucose peptone agar

10 μm

COLONIAL APPEARANCE
at 30°C on glucose peptone agar

colour	cream
surface characteristics	glistening, smooth; sometimes dull, rough

MICROSCOPIC APPEARANCE
at 30°C on cornmeal agar

predominant features	pseudohyphae slow to develop; round to ellipsoidal blastospores

DIFFERENTIAL DIAGNOSIS

Candida glabrata, *C. famata* (not described) and *C. inconspicua* (not described) have similar blastospores, but do not form pseudohyphae and have a different biochemical profile.

SEXUAL STATE

Pichia guilliermondii

CLINICAL IMPORTANCE

It is an occasional cause of superficial and deep-seated infection.

CRYPTOCOCCUS NEOFORMANS

edge of streak on cornmeal agar

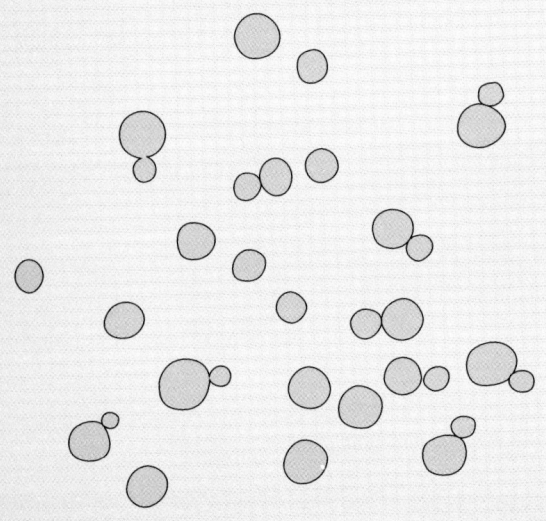

glucose peptone agar

10 μm

COLONIAL APPEARANCE
at 30°C on glucose peptone agar

colour	cream to brownish buff
surface characteristics	dull to mucoid, smooth

MICROSCOPIC APPEARANCE
at 30°C on cornmeal agar

predominant features	large, round blastospores, often well-spaced because of the mucoid capsules

VARIANT FORM

var. *gattii*   antigenically distinct, resistant to canavanine, and assimilates glycine

DIFFERENTIAL DIAGNOSIS

Cryptococcus albidus (not described) and *Cr. laurentii* (not described) have a similar morphological appearance but do not grow at 37°C and have a different biochemical profile; *Cr. albidus* assimilates nitrate.

SEXUAL STATES

Filobasidiella neoformans (of *Cr. neoformans* var. *neoformans*)

Filobasidiella bacillispora (of *Cr. neoformans* var. *gattii*)

CLINICAL IMPORTANCE

It is the cause of cryptococcosis in humans and animals. Infection follows inhalation, but meningitis is the commonest clinical presentation and widespread dissemination can also occur. Cryptococcosis is one of the most common life-threatening fungal infections in patients with AIDS. *Cr. neoformans* var. *neoformans* has a global distribution, but var. *gattii* is most prevalent in Australia and Central Africa. *Cr. neoformans* var. *neoformans* is the predominant cause of cryptococcosis in AIDS patients, even in Africa.

RHODOTORULA GLUTINIS

edge of streak on cornmeal agar

glucose peptone agar

10 μm

COLONIAL APPEARANCE
at 30°C on glucose peptone agar

colour	bright red to orange
surface characteristics	dull, rough to mucoid, smooth

MICROSCOPIC APPEARANCE
at 30°C on cornmeal agar

predominant features	large, round blastospores, sometimes with oval to elongated cells

DIFFERENTIAL DIAGNOSIS

Other *Rhodotorula* spp. have a similar morphological appearance but a different biochemical profile.

SEXUAL STATES

Rhodosporidium diobovatum, R. sphaerocarpum, R. toruloides

CLINICAL IMPORTANCE

It is a common contaminant of skin and nails. It is a rare cause of deep-seated infection in immunocompromised individuals.

SACCHAROMYCES CEREVISIAE

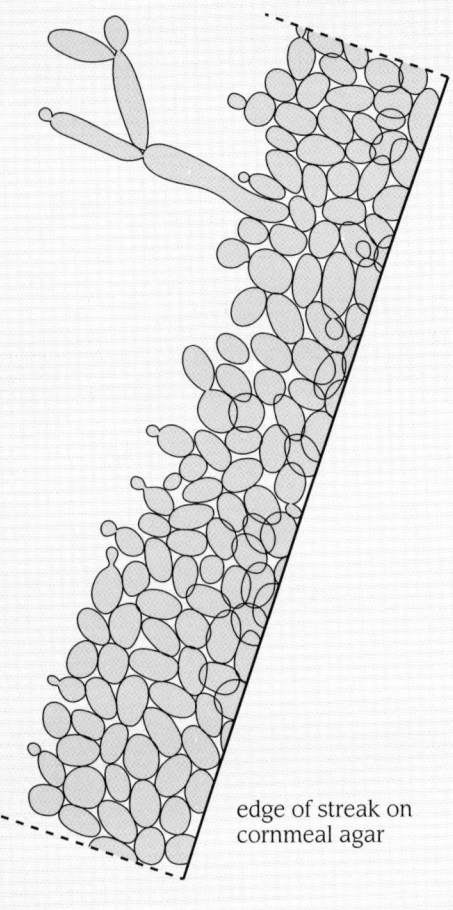

edge of streak on cornmeal agar

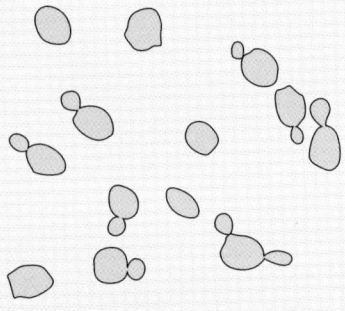

glucose peptone agar

COLONIAL APPEARANCE
at 30°C on glucose peptone agar

colour	cream
surface characteristics	glistening, smooth

MICROSCOPIC APPEARANCE
at 30°C on cornmeal agar

predominant features	large, oval (or elongated) blastospores, sometimes with poorly formed pseudohyphae; ascospores may be found

DIFFERENTIAL DIAGNOSIS

Candida glabrata is similar in morphological appearance but has smaller cells and a different biochemical profile.

CLINICAL IMPORTANCE

This common food organism is an occasional cause of genital infection and a rare cause of deep-seated infection in immunocompromised individuals.

GEOTRICHUM CANDIDUM

COLONIAL APPEARANCE
at 30°C on glucose peptone agar

colour	white to off-white
surface characteristics	flat, spreading, sometimes with a low aerial mycelium

MICROSCOPIC APPEARANCE
at 30°C on cornmeal agar

predominant features	true hyphae, often showing dichotomous branching; large, barrel-shaped arthrospores, resulting from fragmentation of short, lateral branches that arise almost at right angles from the main hyphae; no blastospores

DIFFERENTIAL DIAGNOSIS

Trichosporon spp. and *Blastoschizomyces capitatus* have a similar morphological appearance, but do not show dichotomous branching, do possess blastospores and have a different biochemical profile.

SEXUAL STATE

Galactomyces geotrichum

CLINICAL IMPORTANCE

It is a rare cause of infection in immunocompromised individuals.

BLASTOSCHIZOMYCES CAPITATUS

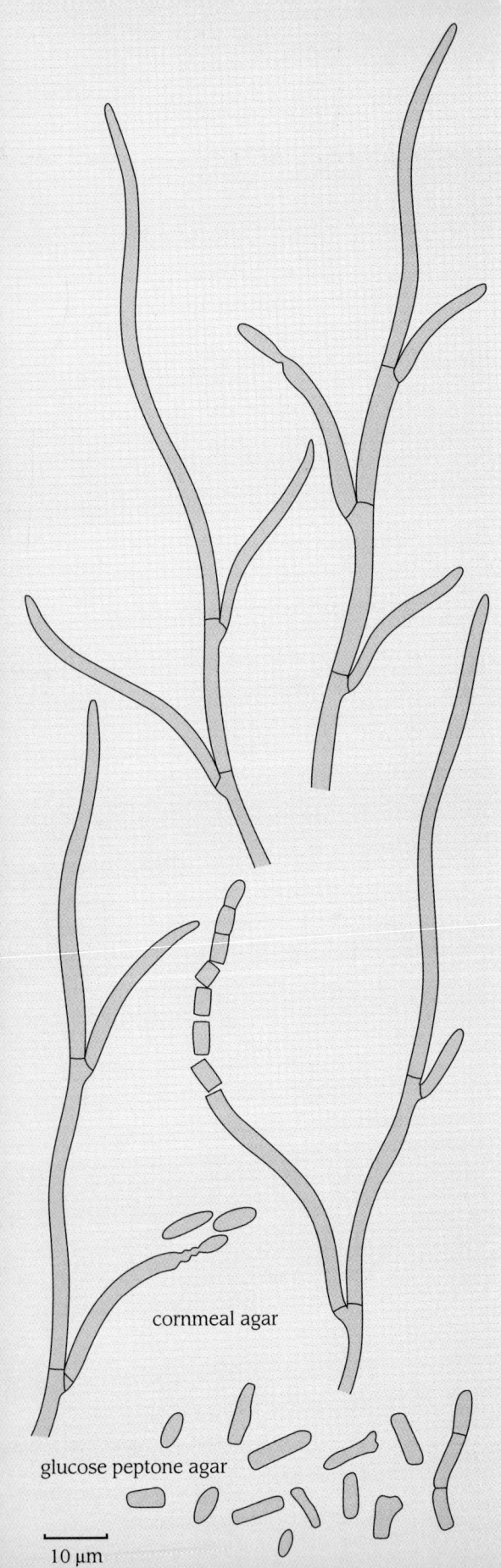

COLONIAL APPEARANCE
at 30°C on glucose peptone agar

colour	white to off-white
surface characteristics	heaped, folded, with a low aerial mycelium

MICROSCOPIC APPEARANCE
at 30°C on cornmeal agar

predominant features	true hyphae, showing characteristic branching with the lateral hyphae curving away from the main stem; branches may fragment into rectangular arthrospores, or produce several blastospores from an annellidic hyphal tip

DIFFERENTIAL DIAGNOSIS

Trichosporon beigelii has a similar morphological appearance, but a different biochemical profile and produces urease.

Geotrichum candidum appears similar, but does not produce blastospores.

SEXUAL STATE

Dipodascus capitatus

CLINICAL IMPORTANCE

It is an occasional cause of deep-seated infection in immunocompromised individuals.

TRICHOSPORON BEIGELII

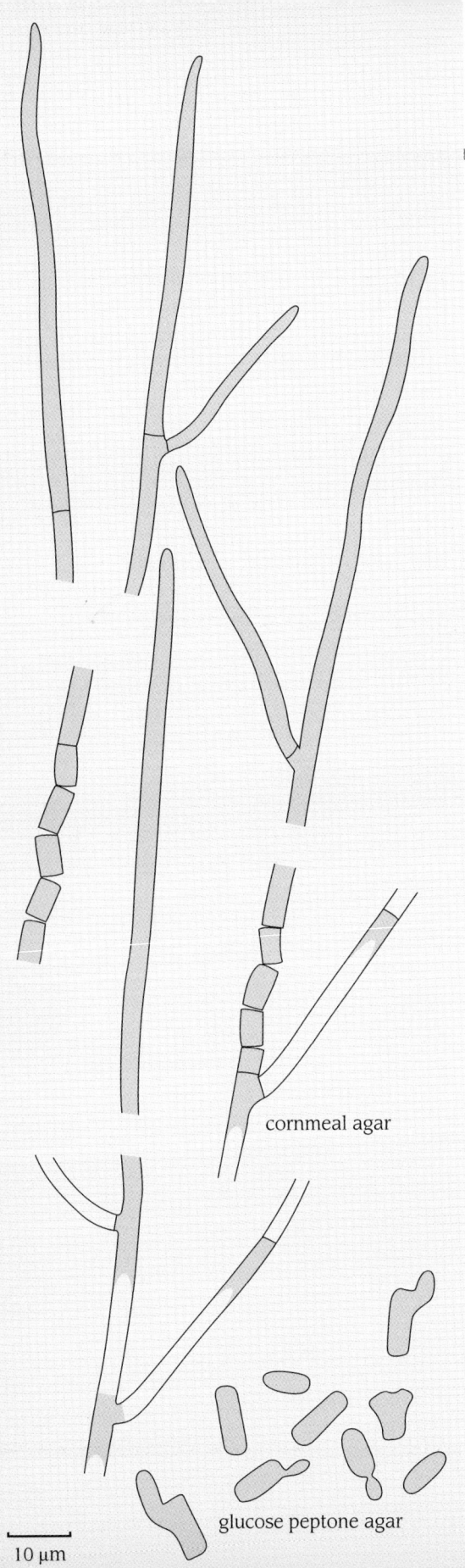

cornmeal agar

glucose peptone agar

10 μm

COLONIAL APPEARANCE
at 30°C on glucose peptone agar

colour	white to cream
surface characteristics	heaped, folded, with a low, white aerial mycelium

MICROSCOPIC APPEARANCE
at 30°C on cornmeal agar

predominant features	true hyphae; cylindrical arthrospores resulting from fragmentation of the older parts of the main stems; blastospores may be present; cytoplasm in older hyphae becomes restricted to the apical end of each cell

DIFFERENTIAL DIAGNOSIS

Geotrichum candidum and *Blastoschizomyces capitatus* have a similar morphological appearance, but a different biochemical profile and do not produce urease.

SEXUAL STATE

None known.

CLINICAL IMPORTANCE

It is the cause of white piedra, a mild infection of the hair. It can also cause localised and disseminated deep-seated infection in debilitated or immunocompromised individuals. The *Trichosporon beigelii* group contains a number of species, named *T. asahii, T. cutaneum, T. inkin, T. mucoides,* and *T. ovoides.*

MALASSEZIA FURFUR

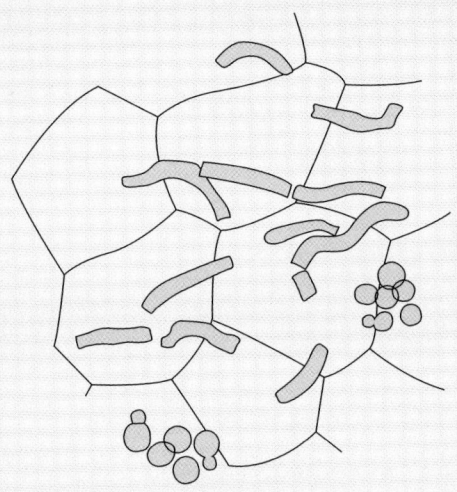

in vivo appearance

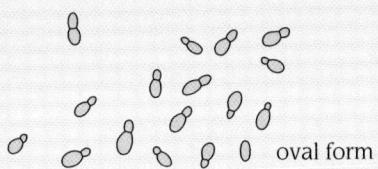

oval form

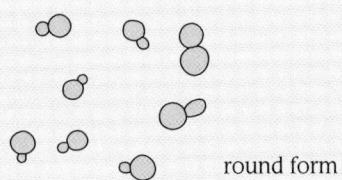

round form

COLONIAL APPEARANCE
at 30°C on glucose peptone agar

No growth or minute colonies.

MICROSCOPIC APPEARANCE
at 30°C on cornmeal agar

No growth.

10 µm

DIFFERENTIAL DIAGNOSIS

Malassezia pachydermatis has similar but larger, oval cells, budding on a broad base.

SEXUAL STATE

None known.

CLINICAL IMPORTANCE

The round-celled form of this lipophilic organism is the cause of the superficial infection, pityriasis versicolor. Microscopic examination of cleared skin scales is sufficient to permit the diagnosis of this condition if round blastospores, together with short, often curved, unbranched hyphae are seen. The oval-celled form is a skin commensal which can cause serious infection in low birthweight infants and debilitated patients receiving parenteral lipid nutrition.

MALASSEZIA PACHYDERMATIS

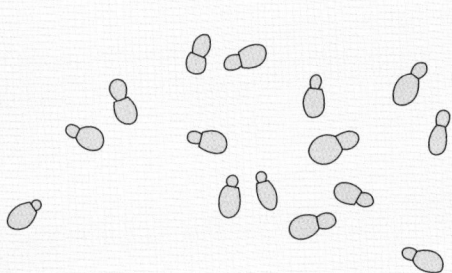

10 µm

COLONIAL APPEARANCE
at 30°C on glucose peptone agar

colour	cream
surface characteristics	small colonies with smooth surface, often having a lobed margin

MICROSCOPIC APPEARANCE
at 30°C on cornmeal agar

Oval cells budding on a broad base.

DIFFERENTIAL DIAGNOSIS

Malassezia furfur

CLINICAL IMPORTANCE

This organism is less fastidious in its requirement for lipid than *M. furfur*. It has caused serious infection in low birthweight infants receiving parenteral nutrition.

APPENDIX 1
COMMON MYCOLOGICAL TERMS

aleuriospore	a thallic conidium that is formed from the end of an undifferentiated hypha, or from a short side-branch.
anamorph	the asexual or imperfect stage of a fungus.
annellide	a specialised conidiogenous cell from which a succession of spores is produced and which has a column of apical scars (annellations) at its tip.
apophysis	the enlargement of a sporangiophore below the columella.
arthrospore	a thallic conidium produced as a result of fragmentation of an existing hypha into separate cells.
ascocarp	a structure that contains asci.
ascoma (pl. -ata)	see **ascocarp**.
ascospore	a haploid spore produced within an ascus following meiosis.
ascus (pl. -ci)	a thin-walled sac containing ascospores, characteristic of the Ascomycotina.
aseptate	without cross-walls or septa.
ballistospore	a conidium that is forcibly discharged.
basidiocarp	a structure that produces basidia.
basidioma (pl. -ata)	see **basidiocarp**.
basidiospore	a haploid spore produced on a basidium following meiosis.
basidium	a cell upon which basidiospores are formed, characteristic of the Basidiomycotina.
basipetal	a succession of conidia, with the youngest at the base of the chain.
blastic	term used to describe the conidia produced as a the result of the enlargement of a part of a conidiogenous cell before a delimiting septum is laid down.
blastospore	a holoblastic conidium produced by the enlargement of a part of a conidiogenous cell before a septum is laid down.

chlamydospore	a resting conidium, formed as a result of the enlargement of an existing hyphal cell.
cleistothecium (pl. -**ia**)	an enclosed ascocarp which splits open to release the ascospores.
columella	swollen tip of the sporangiophore projecting into the sporangium in some Mucorales.
conidiogenesis	term used to describe the developmental processes in the formation of a conidium.
conidiogenous cell	a hypha that produces or becomes a conidium.
conidiophore	the specialised hypha or cell on which, or as part of which, conidia are produced.
conidium (pl. -**ia**)	an asexual, non-motile spore.
dysgonic	slow growing.
enteroblastic	term used to describe the form of conidiogenesis in which a succession of conidia are produced from within a conidiogenous cell.
geniculate	an irregular conidiogenous cell formed by some holoblastic fungi.
gymnothecium (pl. -**ia**)	an ascocarp in which the the asci are distributed within a loose network of hyphae.
heterothallic	a self-sterile fungus; sexual reproduction cannot take place unless two compatible mating strains are present.
hilum	a scar at the base of a conidium.
holoblastic	term used to describe the conidium produced when both the inner and outer walls of the conidiogenous cell swell out to form the conidium.
homothallic	a self-compatible fungus; sexual reproduction can take place within an individual strain.
Hülle cell	large, thick-walled, sterile cells found in some *Aspergillus* spp.

hypha (pl. -ae)	one of the individual filaments that make up the vegetative growth of a fungus.
macroconidium (pl. -ia)	the larger of two different sizes of conidia produced by a fungus in the same manner; it is usually multicellular.
metula (pl. -ae)	apical branch of a conidiophore on which phialides (conidiogenous cells) are borne in *Aspergillus* and *Penicillium* spp.
microconidium (pl. -ia)	the smaller of two different sizes of conidia produced by a fungus in the same manner.
mould	a filamentous fungus.
mycelium	a mass of branching filaments which makes up the vegetative growth of a fungus.
peridial hyphae	the hyphae that make up the outside wall of ascocarps.
perithecium (pl. -ia)	an enclosed ascocarp with an apical opening (ostiole) through which the ascospores are released.
phialide	a specialised conidiogenous cell from which conidia are produced in basipetal succession.
pleomorphic	term used to describe a non-sporing strain of a fungus.
pseudohypha (pl. -ae)	a chain of yeast cells which have arisen as a result of budding and have elongated without becoming detached from each other forming a hypha-like filament.
pseudomycelium	a mass of pseudohyphae.
pycnidiospore	a conidium formed within a pycnidium.
pycnidium (pl. -ia)	an enclosed structure with an apical opening (ostiole) in which conidia are formed.
reflexive	term used to describe a hyphal branch that grows inwards towards the centre of the colony.
rhizoid	a short, branching hypha that resembles a root.
septate	having cross-walls or septa.
septum (pl. -ta)	a cross-wall in a fungal hypha or spore.

sporangiole	a small sporangium.
sporangiophore	a specialised hypha upon which a sporangium develops.
sporangium (pl. **-ia**)	a closed sac containing asexual spores, characteristic of the Zygomycotina.
sporophore	a spore-bearing structure
stroma (pl. **-ata**)	a solid mass of hyphae, sometimes bearing spores on short conidiophores or having ascocarps or pycnidia embedded in it.
synnema (pl. **-ata**)	a compact, elongated cluster of erect conidiophores, their conidia being produced at the tip, along the sides of the upper portion of the synnema, or both.
teleomorph	the sexual or perfect stage of a fungus.
thallic	term used to describe conidia produced as a result of the conversion of an existing hyphal cell.
vesicle	the swollen tip of the conidiophore in *Aspergillus* spp., or swollen part of a sporogenous cell in other fungi.
yeast	a unicellular, budding fungus.
zoospore	a motile asexual spore.
zygospore	a thick-walled, sexual spore produced in the Zygomycotina.

APPENDIX 2
FURTHER READING

COMPREHENSIVE TEXTS

Kibbler CC, Mackenzie DWR, Odds FC (editors). Principles and practice of clinical mycology. Chichester: John Wiley, 1996.

Kwon-Chung KJ, Bennett JE. Medical mycology. 4th ed. Philadelphia: Lea & Febiger, 1992.

Rippon JW. Medical mycology. The pathogenic fungi and the pathogenic actinomycetes. 3rd ed. Philadelphia: Saunders, 1988.

INTRODUCTORY TEXTS

Clayton YM, Midgley G. Medical mycology. London: Gower Medical Publishing, 1985.

Evans EGV, Gentles JC. Essentials of medical mycology. Edinburgh: Churchill Livingstone, 1985.

Richardson MD, Warnock DW. Fungal infection: diagnosis and management. Oxford: Blackwell Science, 1993.

SPECIALISED MONOGRAPHS

Bodey GP (editor). Candidiasis. Pathogenesis, diagnosis and treatment. 2nd ed. New York: Raven Press, 1993.

Elewski BE (editor). Cutaneous fungal infections. New York: Igaku-Shoin, 1991.

Odds FC. Candida and candidosis. 2nd ed. London: Bailliere Tindall, 1988.

Roberts DT, Evans EGV, Allen BR. Fungal nail infections. London: Gower Medical Publishing, 1990.

Ryley JF (editor). Chemotherapy of fungal disease. Berlin: Springer-Verlag, 1990.

Sarosi GA, Davies SF (editors). Fungal diseases of the lungs. 2nd ed. New York: Raven Press, 1993.

Smith JMB. Opportunistic mycoses of man and animals. Wallingford: CAB International, 1989.

Warnock DW, Richardson MD (editors). Fungal infection in the compromised patient. 2nd ed. Chichester: John Wiley, 1991.

LABORATORY DIAGNOSIS OF FUNGAL INFECTION

Evans EGV, Richardson MD (editors). Medical mycology: a practical approach. Oxford: IRL Press at Oxford University Press, 1989.

Koneman EW, Roberts GD. Practical laboratory mycology. 3rd ed. Baltimore: Williams and Wilkins, 1985.

McGinnis MR. Laboratory handbook of medical mycology. New York: Academic Press, 1980.

IDENTIFICATION MANUALS

Barnett JA, Payne RW, Yarrow D. Yeasts: characteristics and identification. 2nd ed. Cambridge University Press, 1990.

De Hoog GS, Guarro J (editors). Atlas of clinical fungi. Baarn: Centraalbureau voor Schimmelcultures, 1995.

Kreger-van Rij NJW (editor). The yeasts. 3rd ed. Amsterdam: Elsevier Science Publishers, 1984.

Larone DH. Medically important fungi. A guide to identification. 3rd ed. Washington DC: ASM Press, 1995.

St Germain G, Summerbell R. Identifying filamentous fungi. A clinical laboratory handbook. Belmont: Star Publishing, 1995.

INDEX

A

Absidia spp. 197, 201, 203, 211

Absidia corymbifera 188, **192-3**

Absidia ramosa, see *Absidia corymbifera*

Acremonium spp. 7, 146, 147, 148, 153, 155, 157, 159, 161, 163, 165, 173

Acremonium kiliense 147, 148, **164-5**

Acremonium strictum 147, **162-3**

Ajellomyces capsulatus 81

Ajellomyces dermatitidis 83

Allescheria boydii, see *Scedosporium apiospermum*

Alternaria spp. 109, 111, 113, 115, 145

Alternaria alternata 88, **106-7**

Aphanoascus fulvescens 77, 217, **218-19**

Aphanoascus keratinophilus 77

Apophysomyces elegans 189, **210-11**

Arachnomyces nodososetosus 25

Arthrinium spp. 21, 169, 171

Arthroderma benhamiae 47, 49, 51

Arthroderma fulvum 37

Arthroderma gypseum 37

Arthroderma incurvatum 37

Arthroderma insingulare 41

Arthroderma lenticularum 41

Arthroderma otae 33

Arthroderma persicolor 43

Arthroderma quadrifidum 41

Arthroderma vanbreuseghemii 49, 51

Aspergillus spp. 7, 12, 116, 117, 118, 135, 139, 219

Aspergillus candidus 119, 133, **136-7**

Aspergillus clavatus 118, 123

Aspergillus flavipes 119

Aspergillus flavus 119, **120-21**

Aspergillus fumigatus 119, **122-3**, 125, 127

Aspergillus glaucus 119, 123, **124-5**, 127, 216

Aspergillus nidulans 119, 123, 125, **126-7**, 129, 131, 216

Aspergillus niger 119, **132-3**, 137

Aspergillus ochraceus 119, 121, 133, 137

Aspergillus oryzae 121

Aspergillus parasiticus 121

Aspergillus terreus 79, 119, **134-5**, 141, 145

Aspergillus ustus 119, 129, **130-31**, 133, 171

Aspergillus versicolor 119, 123, 125, 127, **128-9**, 131

Aureobasidium spp. 181

Aureobasidium pullulans 87, **90-91**, 93, 167, 169, 179

B

Basidiobolus spp. 186, 207

Basidiobolus haptosporus, see *Basidiobolus ranarum*

Basidiobolus meristosporus, see *Basidiobolus ranarum*

Basidiobolus ranarum 188, **204-5**, 209

Basipetospora rubra 221

Bipolaris spp. 109, 111, 113, 115

Bipolaris australiensis 88, **113**

Bipolaris hawaiiensis 88, **112-13**

Blastomyces dermatitidis 23, 72, 73, 81, **82-3**, 85, 139

Blastoschizomyces capitatus 243, 245, 246, 249, 277, **278-9**, 281

Botryodiplodia theobromae, see *Lasiodiplodia theobromae*

C

Candida spp. 243, 244, 249, 265

Candida albicans 242, 243, 244, 245, 246, 247, 249, **250-51**, 253, 257

Candida dubliniensis 251

Candida famata 267, 269

Candida glabrata 245, 246, **266-7**, 269, 275

Candida guilliermondii 244, 245, 246, 261, 263, **268-9**

Candida inconspicua 267, 269

Candida kefyr 245, 246, 255, **258-9**, 261, 263

Candida krusei 245, 246, **254-5**, 259

Candida lipolytica 245, 246, **256-257**

Candida lusitaniae 245, 246, **260-61**

Candida parapsilosis 245, 246, 261, **262-63**

Candida pelliculosa 245, 246, **264-5**

Candida tropicalis 243, 245, 246, 251, **252-3**, 257

Chaetomium spp. 221, 223

Chaetomium globosum 217, **222-3**

Chrysosporium spp. 21, 23, 25, 26, 49, 51, 72, 75, 77, 79, 81, 219

Chrysosporium keratinophilum 39, 72, 73, **76-7**, 218, 219

Chrysosporium pannorum, see *Geomyces pannorum*

Cladophialophora spp. 95, 97, 239

Cladophialophora bantiana 87, **95**

Cladophialophora carrionii 87, **94-5**

Cladosporium spp. 6, 95, 97, 99, 101, 103, 175, 177, 181, 183, 185, 231

Cladosporium bantianum, see *Cladophialophora bantiana*

Cladosporium carrionii, see *Cladophialophora carrionii*

Cladosporium cladosporioides 87, **97**

Cladosporium herbarum 87, **97**

Cladosporium sphaerospermum 87, **96-7**

Cladosporium trichoides, see *Cladophialophora bantiana*

Cladosporium werneckii, see *Phaeoannellomyces werneckii*

Clavispora lusitaniae 261

Coccidioides immitis 16, 17, **22-3**, 25

Cochliobolus lunatus 111

Cochliobolus hawaiiensis 113

Conidiobolus spp. 186

Conidiobolus coronatus 188, 205, **206-7**, 209

Cryptococcus spp. 81, 83, 85, 244, 247

Cryptococcus albidus 271

Cryptococcus laurentii 271

Cryptococcus neoformans 242, 243, 244, 245, 246, 247, 249, **270-71**

Cunninghamella spp. 186

Cunninghamella bertholletiae 188, **190-91**

Cunninghamella elegans 191

Curvularia spp. 107, 109, 113, 115

Curvularia lunata 88, **110-11**

Cylindrocarpon spp. 146, 151, 159, 161

Cylindrocarpon lichenicola 147, **150-51**

D

Dipodascus capitatus 279

Drechslera spp. 113

Drechslera australiensis, see *Bipolaris australiensis*

Drechslera hawaiiensis, see *Bipolaris hawaiiensis*

Drechslera rostrata, see *Exserohilum rostratum*

E

Emericella nidulans 127

Emmonsia parva 83

Entomophthora coronata, see *Conidiobolus coronatus*

Epidermophyton floccosum 27, 28, 29, **38-9**, 55, 57, 77

Eurotium spp. 125

Exophiala spp. 91, 95, 97, 99, 101, 103, 148, 169, 175, 177, 179, 181, 183, 185, 231, 233

Exophiala dermatitidis 149, 179, **182-3**, 185

Exophiala jeanselmei 149, 179, 183, **184-5**

Exophiala spinifera 149, **180-81**

Exophiala werneckii, see *Phaeoannellomyces werneckii*

Exserohilum spp. 111, 113, 115

Exserohilum longirostratum 88, **115**

Exserohilum mcginnisii 88, **115**

Exserohilum rostratum 88, **114-15**

F

Filobasidiella bacillispora 271

Filobasidiella neoformans 271

Fonsecaea spp. 95, 97, 101, 175, 177, 183, 185

Fonsecaea pedrosoi 87, **98-9**, 101, 103

Fusarium spp. 7, 143, 146, 147, 151, 153, 155, 157, 159, 161, 163, 165, 221

Fusarium dimerum 147, **152-3**

Fusarium moniliforme 147, 148, **156-7**, 159

Fusarium oxysporum 147, **158-9**, 161

Fusarium semitectum 147, **154-5**

Fusarium solani 147, 159, **160-61**

G

Galactomyces geotrichum 277

Geomyces pannorum 72, 73, **74-5**

Geotrichum spp. 17

Geotrichum candidum 16, 21, 243, 245, 246, 249, **276-7**, 279, 281

Geotrichum capitatum, see *Blastoschizomyces capitatus*

H

Hansenula anomala,
see *Candida pelliculosa*

Helminthosporium spp. 113, 115

Hendersonula toruloidea,
see *Scytalidium dimidiatum*

Histoplasma capsulatum 72, 73, 79, **80-81**, 83

Histoplasma capsulatum var. *duboisii* **81**, 83, 85

Hyphomyces destruens,
see *Pythium insidiosum*

I

Issatchenkia orientalis 255

K

Kluyveromyces marxianus 259

L

Lasiodiplodia theobromae 19, 217, **236-7**

Leptosphaeria senegalensis 217, **230-31**

Lecythophora spp. 91, 147, 148, 169, 173

Lecythophora hoffmannii **167**, 225

Lecythophora mutabilis 93, 147, **166-7**

M

Madurella spp. 216, 233, 235

Madurella grisea 217, 239, **241**

Madurella mycetomatis 217, **240-41**

Malassezia furfur 249, **282-3**, 285

Malassezia pachydermatis 249, 283, **284-5**

Malbranchea spp. 17, 23, 25

Microsporum spp. 33

Microsporum amazonicum **68**

Microsporum audouinii 27, 28, 31, 33, 35, **58-9**

Microsporum boullardii **68**

Microsporum canis 27, 28, 29, 31, **32-3**, 35, 47

Microsporum canis dysgonic form **33**, 57, 59, 61

Microsporum cookei **69**

Microsporum distortum 33, **69**

Microsporum equinum 27, 29, 33, **34-5**, 59

Microsporum fulvum 27, 29, **37**

Microsporum gallinae **70**

Microsporum gypseum 27, 28, 29, **36-7**

Microsporum nanum **69**

Microsporum persicolor 27, 28, 30, **42-3**, 45

Microsporum racemosum **68**

Microsporum vanbreuseghemii **69**

Monascus ruber 141, 143, 217, **220-21**

Monosporium apiospermum,
see *Scedosporium apiospermum*

Mortierella wolfii 189, **214-15**

Mucor spp. 193, 195, 197, 201, 203, 211

Mucor circinelloides 188, **196-7**

Mucor hiemalis 188, **202-3**

Mucor pusillus, see *Rhizomucor pusillus*

Mucor racemosus 188, **203**

Myceliophthora spp. 72, 81

Myceliophthora thermophila 72, 73, 77, **78-9**, 135, 141

Myxotrichum deflexum 217, **226-7**

N

Nattrassia mangiferae 18

Neotestudina rosatii 217, **232-3**

O

Ochroconis gallopava 88, **104-5**, 221

Onychocola canadensis 17, **24-5**

P

Paecilomyces spp. 143, 145, 148, 157

Paecilomyces lilacinus 119, **142-3**, 159, 161

Paecilomyces variotii 119, 135, **144-5**

Paracoccidioides brasiliensis 23, 72, 73, 83, **84-5**

Penicillium spp. 7, 12, 116, 117, 118, 119, 123, 125, 127, 129, 139, 141, 143, 145

Penicillium marneffei 118, **138-9**, 227

Petriellidium boydii,
see *Scedosporium apiospermum*

Phaeoannellomyces werneckii 149, **178-9**

Phialemonium spp. 147, **167**

Phialophora spp. 95, 97, 99, 101, 103, 145, 148, 167, 175, 177, 181, 183, 185, 231, 233, 239

Phialophora hoffmannii,
see *Lecytophora hoffmannii*

Phialophora jeanselmei,
see *Exophiala jeanselmei*

Phialophora mutabilis,
see *Lecythophora mutabilis*

Phialophora parasitica 147, 149, **172-3**, 175

Phialophora pedrosoi,
see *Fonsecaea pedrosoi*

Phialophora richardsiae 149, 173, **174-5**

Phialophora spinifera,
see *Exophiala spinifera*

Phialophora verrucosa 148, 149, 175, **176-7**

Phoma spp. 225, 239

Phoma herbarum 217, 224-225

Pichia anomala 265

Pichia guilliermondii 269

Piedraia hortae 217, **234-5**

Pityrosporum ovale, see *Malassezia furfur*

Pityrosporum orbiculare,
see *Malassezia furfur*

Pityrosporum pachydermatis,
see *Malassezia pachydermatis*

Pseudallescheria boydii 171

Pullularia pullulans,
see *Aureobasidium pullulans*

Pyrenochaeta spp. 223

Pyrenochaeta romeroi 217, **238-9**, 241

Pyrenochaeta unguis-hominis **239**

Pythium spp. 189

Pythium insidiosum 187, **208-9**

R

Ramichloridium spp. 103

Ramichloridium mackenziei 87, **102-3**

Renispora spp. 81

Rhinocladiella spp. 99, 101, 103

Rhinocladiella atrovirens 87, **100-101**

Rhinocladiella pedrosoi,
see *Fonsecaea pedrosoi*

Rhizomucor spp. 193, 197, 201, 211

Rhizomucor pusillus 188, **194-5**

Rhizopus spp. 193, 197, 201, 211

Rhizopus arrhizus 188, 199, **200-201**

Rhizopus microsporus 188, **198-9**

Rhizopus nigricans,
see *Rhizopus stolonifer*

Rhizopus oligosporus,
see *Rhizopus microsporus*

Rhizopus oryzae, see *Rhizopus arrhizus*

Rhizopus rhizopodiformis,
see *Rhizopus microsporus*

Rhizopus stolonifer 188, **201**

Rhodosporidium diobovatum 273

Rhodosporidium sphaerocarpum 273

Rhodosporidium toruloides 273

Rhodotorula spp. 244, 247, 249, 273

Rhodotorula glutinis 245, 246, **272-3**

S

Saccharomyces cerevisiae 245, 246, 267, **274-5**

Saksenaea vasiformis 189, **212-13**

Scedosporium spp. 8, 77, 147, 171, 173

Scedosporium apiospermum 131, 147, 169, **170-71**, 216

Scedosporium inflatum,
see *Scedosporium prolificans*

Scedosporium prolificans 147, 167, **168-9**, 181, 183, 185

Schizophyllum commune 217, **228-9**

Scopulariopsis spp. 8, 141

Scopulariopsis brevicaulis 39, 79, 119, 135, **140-41**, 145, 221

Scytalidium dimidiatum 16, 17, **18-19**, 20, 21, 216, 237

Scytalidium hyalinum 17, 19, **20-21**

Sepedonium spp. 81

Setosphaeria rostrata 115

Sporothrix spp. 93

Sporothrix schenckii 87, 91, **92-3**, 167, 169, 173

T

Torulopsis glabrata, see *Candida glabrata*

Trichophyton spp. 28, 33, 219

Trichophyton ajelloi 29, 41, **71**

Trichophyton concentricum 27, 28, 31, 33, 65, **66-7**

Trichophyton erinacei 27, 30, 33, 45, **46-7**, 49, 51, 53

Trichophyton equinum 27, 30, 43, **44-5**, 47, 49, 51

Trichophyton gourvilii 65, **71**

Trichophyton interdigitale 13, 27, 30, 41, 43, 45, 47, 49, **50-51**, 53

Trichophyton interdigitale nodular form 30, 31, 47, **51**, 57

Trichophyton megninii 71

Trichophyton mentagrophytes 27, 28, 30, 41, 43, 45, 47, **48-9**, 51, 53, 55, 75, 77

Trichophyton rubrum 13, 27, 30, 31, 43, 45, 47, 51, **52-3**, 63, 65

Trichophyton rubrum granular form 29, 30, 49, 51, **53**, 55, 57

Trichophyton schoenleinii 27, 28, 31, 33, **60-61**, 65, 67, 83

Trichophyton simii 70

Trichophyton soudanense 27, 30, 33, 39, 53, 55, **56-7**

Trichophyton terrestre 29, **40-41**, 49, 51, 137

Trichophyton tonsurans 27, 28, 30, 39, 49, 53, **54-5**, 57, 77

Trichophyton verrucosum 27, 28, 31, 33, 61, **62-3**, 65

Trichophyton violaceum 27, 28, 31, 39, 53, 57, 61, **64-5**

Trichophyton yaoundei 65, **71**

Trichosporon spp. 17, 81, 83, 85, 93, 139, 167, 243, 244, 247, 249, 277

Trichosporon asahii 281

Trichosporon beigelii 243, 244, 245, 246, 249, 279, **280-81**

Trichosporon capitatum,
see *Blastoschizomyces capitatus*

Trichosporon cutaneum 281

Trichosporon inkin 281

Trichosporon mucoides 281

Trichosporon ovoides 281

U

Ulocladium spp. 6, 107, 111, 113, 115

Ulocladium chartarum 88, **108-9**

W

Wangiella spp. 181, 185

Wangiella dermatitidis,
see *Exophiala dermatitidis*

Y

Yarrowia lipolytica 257

X

Xylohypha bantiana,
see *Cladophialophora bantiana*